RENEWABLE ENERGY: RESEARCH, DEVELOPMENT AND POLICIES

RETHINKING NUCLEAR POWER IN THE UNITED STATES

KENNETH A. VELLIS
EDITOR

Nova Science Publishers, Inc.
New York

For permission to use material from this book please contact us:
Telephone 631-231-7269; Fax 631-231-8175
Web Site: http://www.novapublishers.com

LIBRARY OF CONGRESS CATALOGING-IN-PUBLICATION DATA

Rethinking nuclear power in the United States / editor, Kenneth A. Vellis.
p. cm.
Includes index.
ISBN 978-1-60692-152-4 (hardcover)
1. Nuclear energy--Government policy--United States. 2. Radioactive waste disposal--Government policy--United States. 3. Nuclear power plants--United States--Safety measures. I. Vellis, Kenneth A.
TK9023.R473 2009
333.792'40973--dc22

2009041003

Published by Nova Science Publishers, Inc. ✝ New York

CONTENTS

PREFACE

This book explores the topic of nuclear power in the United States. Nearly three decades after the most recent order was placed for a new nuclear power plant in the U.S., several utilities are now expressing interest in building a total of up to 30 new reactors. The renewed interest in nuclear power has resulted primarily from higher prices for natural gas, improved operation of existing reactors, and uncertainty about future restrictions on coal emissions. This book compares the cost of two-fuel cycle alternatives for the current generation of thermal reactors - one alternative being direct disposal, and the other reprocessing. This book also includes analyses of the potential effect of the tax credit for nuclear power provided by the Energy Policy of 2005 and possible competitive effects of various proposals to limit greenhouse gas emissions. Under baseline assumptions, the cost of electricity from new nuclear power plants is likely to be higher than power generated by new coal- and natural gas-fired plants. This book consists of public documents which have been located, gathered, combined, reformatted, and enhanced with a subject index, selectively edited and bound to provide easy access.

Chapter 1 - By the end of the next decade, demand for electricity in the United States is expected to increase by about 20 percent, according to the Energy Information Administration (EIA). That projected increase—coupled with concerns about the effects of greenhouse-gas emissions on the environment—has encouraged policymakers to reassess the role that nuclear power might play both in expanding the capacity to generate electricity and in limiting the amount of greenhouse gases produced by the combustion of fossil fuels. Because nuclear power uses an abundant fuel source to generate electricity without emitting such gases, prospects that new nuclear power plants will be planned and financed in the next decade are greater than at any time since the 1970s, when cost overruns and concerns about public safety halted investment in such facilities.

This reappraisal of nuclear power is motivated in large part by the expectation that market-based approaches to limit greenhouse-gas emissions could be put in place in the near future. Several options currently being considered by the Congress—including "cap-and-trade" programs—would impose a price on emissions of carbon dioxide, the most common greenhouse gas.[1] If implemented, such limits would encourage the use of nuclear technology by increasing the cost of generating electricity with conventional fossil-fuel technologies. The prospect that such legislation will be enacted is probably already reducing investment in conventional coal-fired power plants.

Chapter 2 - Nearly three decades after the most recent order was placed for a new nuclear power plant in the United States, several utilities are now expressing interest in building a total of up to 30 new reactors. The renewed interest in nuclear power has resulted primarily from higher prices for natural gas, improved operation of existing reactors, and uncertainty about future restrictions on coal emissions. A substantial tax credit and other incentives for nuclear generation provided by the Energy Policy Act of 2005 (P.L. 109-58) are also likely to improve the economic viability of qualifying new reactors. New nuclear plant applications can also take advantage of amendments to the Atomic Energy Act made in the early 1990s to reduce licensing delays.

Currently, there are 103 licensed and operable power reactors at 65 plant sites in 31 states, generating about one-fifth of U.S. electricity. Although no new U.S. reactors have started up since 1996, U.S. nuclear electricity generation has since grown by more than 20%. Much of this additional output resulted from reduced downtime, notably through shorter refueling outages. Licensed commercial reactors generated electricity at an average of 89.8% of their total capacity in 2006, after averaging about 75% in the mid-1990s and about 65% in the mid-1980s.

Falling operating costs have helped renew the economic viability of the nation's fleet of nuclear power plants. From 1989 to 1998, 12 commercial reactors were closed before reaching the end of their 40-year licenses. By the late 1990s, there was real doubt that any reactors would make it to 40 years. Since 2000, however, 44 commercial reactors have received 20-year license extensions from the Nuclear Regulatory Commission (NRC), giving them up to 60 years of operation, and more are pending.

The nuclear production tax credit in the Energy Policy Act could have a significant impact on the economic viability of new nuclear power plants. Under base case assumptions, new reactors are not competitive with either coal-fired or natural gas-fired facilities. However, if new reactors are able to take full advantage of the nuclear production tax credit, nuclear power appears competitive with either natural gas-fired or coal-fired facilities.

Other factors will also be important in the commercial decision to invest in new nuclear plants, such as fossil fuel prices and the regulatory environment for both nuclear power and future fossil fuel-fired generation. If natural gas prices remain at historically high levels, future nuclear plants will be more likely to be competitive without federal tax credits. However, natural gas prices have been highly cyclical in the past, raising the possibility that nuclear costs could be undercut in the future.

Any substantial mandatory greenhouse gas control program would probably affect the cost of new coal-fired and natural gas-fired generation relative to nuclear power, particularly if nuclear power is assumed to have no greenhouse gas emissions. Continued delays in nuclear waste disposal facilities — forcing spent fuel to be stored at plant sites — could also affect the decision to construct new reactors.

Chapter 3 – This is a testimony of Peter R. Orszag, Director, Congressional Budget Office, before the Committee on Energy and Natural Resources, United States Senate.

Chapter 4 – This is a testimony of Kim Cawley, Chief, Natural and Physical Resources, Cost Estimates Unit, before the Committee on the Budget, United States House of Representatives.

Chapter 5 - Protection of nuclear power plants from land-based assaults, deliberate aircraft crashes, and other terrorist acts has been a heightened national priority since the attacks of September 11, 2001. The Nuclear Regulatory Commission (NRC) has strengthened its regulations on nuclear reactor security, butcritics contend that implementation by the industry has been too slow and that further measures are needed. Several provisions to increase nuclear reactor security were included in the Energy Policy Act of 2005, signed August 8, 2005. The law requires NRC to conduct "force-on-force" security exercises at nuclear power plants at least once every three years and to revise the "design-basis threat" that nuclear plant security forces must be able to meet, among other measures.

Chapter 6 - Spent nuclear fuel and high-level radioactive waste are materials from nuclear power plants and government defense programs. These materials contain highly radioactive elements, such as cesium, strontium, technetium, and neptunium. Some of these elements will remain radioactive for a few years, while others will be radioactive for millions of years.

Scientists worldwide agree that the safest way to manage these materials is to dispose of them deep underground in what is called a geologic repository.

Chapter 7 - Some of the electricity used in homes across our nation is made with nuclear fuel in reactors at power plants. When the fuel that is used to make electricity at the power plants can no longer produce electricity efficiently, it is removed from the reactors. This used fuel is called spent fuel. Some nuclear materials used for national defense create waste. This waste is called high-level radioactive waste. These highly radioactive materials must be protected so they won't harm people or the environment.

In: Rethinking Nuclear Power in the United States ISBN: 978-1-60692-152-4
Editor: Kenneth A. Vellis

Chapter 1

NUCLEAR POWER'S ROLE IN GENERATING ELECTRICITY*

Congressional Budget Office

1. SUMMARY AND INTRODUCTION

By the end of the next decade, demand for electricity in the United States is expected to increase by about 20 percent, according to the Energy Information Administration (EIA). That projected increase—coupled with concerns about the effects of greenhouse-gas emissions on the environment—has encouraged policymakers to reassess the role that nuclear power might play both in expanding the capacity to generate electricity and in limiting the amount of greenhouse gases produced by the combustion of fossil fuels. Because nuclear power uses an abundant fuel source to generate electricity without emitting such gases, prospects that new nuclear power plants will be planned and financed in the next decade are greater than at any time since the 1970s, when cost overruns and concerns about public safety halted investment in such facilities.

This reappraisal of nuclear power is motivated in large part by the expectation that market-based approaches to limit greenhouse-gas emissions

* This is an edited, reformatted and augmented version of a Congressional Budget Office publication dated May 2008.

could be put in place in the near future. Several options currently being considered by the Congress—including "cap-and-trade" programs—would impose a price on emissions of carbon dioxide, the most common greenhouse gas.[1] If implemented, such limits would encourage the use of nuclear technology by increasing the cost of generating electricity with conventional fossil-fuel technologies. The prospect that such legislation will be enacted is probably already reducing investment in conventional coal-fired power plants.

Current energy policy, especially as established and expanded under the Energy Policy Act of 2005 (EPAct), provides incentives for building additional capacity to generate electricity using innovative fossil-fuel technologies and an advanced generation of nuclear reactor designs that are intended to decrease costs and improve safety.[2] Among the provisions of EPAct that specifically apply to newly built nuclear power plants are funding for research and development; investment incentives, such as loan guarantees and insurance against regulatory delays; and production incentives, including a tax credit. Since the enactment of EPAct, about a dozen utilities have announced their intention to license about 30 nuclear plants.

This study assesses the commercial viability of advanced nuclear technology as a means of meeting future demand for electricity by comparing the costs of producing electricity from different sources under varying circumstances. The Congressional Budget Office (CBO) estimated the cost of producing electricity using a new generation of nuclear reactors and other base-load technologies under a variety of assumptions about prospective carbon dioxide charges, EPAct incentives, and future market conditions.[3] This study compares the cost of advanced nuclear technology with that of other major sources of base-load capacity that are available throughout the country—including both conventional and innovative fossil-fuel technologies. Because the study focuses only on technologies that can be used as base-load capacity in most parts of the country, it does not address renewable energy technologies that are intermittent (such as wind and solar power) or technologies that use resources readily available only in certain areas (such as geothermal or hydroelectric power).

In the long run, carbon dioxide charges would increase the competitiveness of nuclear technology and could make it the least expensive source of new base-load capacity. More immediately, EPAct incentives by themselves could make advanced nuclear reactors a competitive technology for limited additions to base-load capacity. However, under some plausible assumptions that differ from those CBO adopted for its reference scenario—in particular, those that project higher future construction costs for nuclear plants or lower natural gas prices—nuclear technology would be a relatively expensive source of capacity, regardless of EPAct incentives. CBO's analysis yields the following conclusions:

- In the absence of both carbon dioxide charges and EPAct incentives, conventional fossil-fuel technologies would most likely be the least expensive source of new electricity-generating capacity.
- Carbon dioxide charges of about \$45 per metric ton would probably make nuclear generation competitive with conventional fossil-fuel technologies as a source of new capacity, even without EPAct incentives. At charges below that threshold, conventional gas technology would probably be a more economic source of base-load capacity than coal technology. Below about \$5 per metric ton, conventional coal technology would probably be the lowest cost source of new capacity.
- Also at roughly \$45 per metric ton, carbon dioxide charges would probably make nuclear generation competitive with existing coal power plants and could lead utilities in a position to do so to build new nuclear plants that would eventually replace existing coal power plants.
- EPAct incentives would probably make nuclear generation a competitive technology for limited additions to base-load capacity, even in the absence of carbon dioxide charges. However, because some of those incentives are backed by a fixed amount of funding, they would be diluted as the number of nuclear projects increased; consequently, CBO anticipates that only a few of the 30 plants currently being proposed would be built if utilities did not expect carbon dioxide charges to be imposed.
- Uncertainties about future construction costs or natural gas prices could deter investment in nuclear power. In particular, if construction costs for new nuclear power plants proved to be as high as the average cost of nuclear plants built in the 1970s and 1980s or if natural gas prices fell back to the levels seen in the 1990s, then new nuclear capacity would not be competitive, regardless of the incentives provided by EPAct. Such variations in construction or fuel costs would be less likely to deter investment in new nuclear capacity if investors anticipated a carbon dioxide charge, but those charges would probably have to exceed \$80 per metric ton in order for nuclear technology to remain competitive under either of those circumstances.

Background on Electricity-Generating Technologies

Electricity is produced using a variety of technologies powered by different sources of fuel, but the sources that predominate are coal, natural gas, and uranium. Coal-burning technologies emit the most carbon dioxide per unit of electricity; natural gas technologies emit carbon dioxide at about half that rate; and nuclear power, a "zero emissions technology," emits no carbon dioxide at all.[4] (See Box 1-1 for details on power plant technologies.)

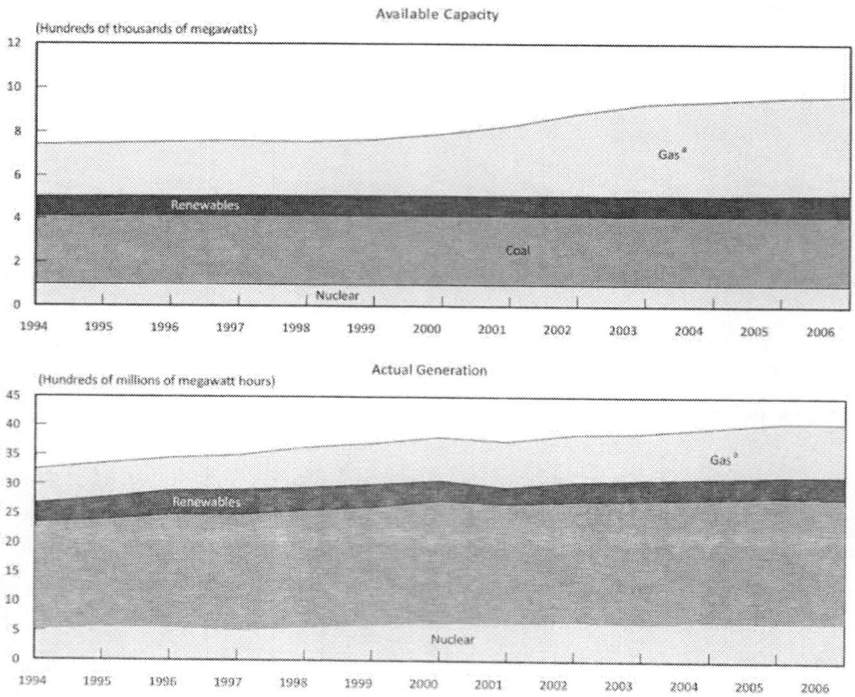

Source: Congressional Budget Office based on data from the Energy Information Administration.

Notes: Electricity-generating capacity is measured in megawatts; the electrical power generated by that capacity is measured in megawatt hours. During a full hour of operation, 1 megawatt of capacity produces 1 megawatt hour of electricity, which can power roughly 800 average households.

a Natural gas capacity represents about 85 percent of total gas capacity, with the remainder using petroleum-based or other gases

Figure 1-1. Available Capacity to Generate Electricity and Actual Generation, by Technology Type, 1994 to 2006

Historically, most base-load capacity has been provided using coal or nuclear technologies because, once the plants have been built, low fuel costs make them relatively cheap to operate continuously. Even though natural gas prices have increased significantly in recent years, natural gas remains the dominant source of peak capacity because power plants using that fuel are less expensive to build than coal-fired plants or nuclear reactors and easier to start up and shut down.

From 1994 to 2006, a period when the total amount of electricity generated rose by 25 percent, utilities appear to have increased base-load generation primarily by stepping up production at existing coal and nuclear plants, while using natural gas technology to expand overall capacity. During that time, the amount of electricity generated by nuclear and coal power plants expanded by about 20 percent as those facilities operated at closer to maximum capacity. (See Figure 1-1.) However, coal and nuclear capacity remained roughly constant because utilities increasingly chose to construct new natural gas power plants. By 2006, natural gas capacity had doubled, accounting for 75 percent more electricity generation than in 1994. Because of the extensive investment in natural gas capacity, the amount of excess capacity—which is used to meet surges in demand or to compensate for shutdowns at power plants—remained roughly the same.[5]

Box 1-1. Technologies for Adding to Base-Load Capacity over the Next Decade

An advanced generation of nuclear reactors is one of several options currently under consideration for providing additional base-load capacity to produce electricity. Utilities will weigh the cost of new nuclear power plants against that of both conventional and innovative fossil-fuel alternatives. Among the conventional alternatives are pulverized coal technology and combined-cycle turbines that rely on natural gas. Among the innovative alternatives are technologies that capture and store most of the carbon dioxide emitted when coal and natural gas are burned. Utilities use several other technologies to generate electricity, but they are not widespread and are not likely to be commercially viable base-load alternatives in most areas.

Nuclear Power Plants

Advanced, or third-generation, nuclear reactors were developed by enhancing the designs of existing nuclear power plants, which use first-and second-generation reactors developed before the 1980s, before major advances in digital control systems. Interest in those older designs disappeared in the 1970s for a variety of reasons, including construction cost overruns, poor operational performance, and concerns about the safety of the nuclear technology. Beginning in the 1990s, industry and government participated in a variety of cost-sharing programs to develop the third generation of nuclear reactors, which are designed to be safer to operate and less expensive to build and maintain.[1]

Fossil-Fuel Power Plants

Coal and natural gas can be burned to create electricity through several different technologies. Pulverized coal power plants, which burn solid coal ignited by injected air, are by far the most common option for generating base-load electricity. Combined-cycle technology, which harnesses residual steam heat from the combustion cycle, has become, over the past 20 years, the most efficient method of generating electricity from natural gas. However, those conventional coal and natural gas technologies emit carbon dioxide and are therefore susceptible to the pricing of such emissions, so innovative technologies with "carbon capture and storage" (CCS) may become the most commercially viable option for using those fossil fuels.

Although CCS technologies have not yet been deployed commercially at power plants, many observers expect them to be available over the upcoming decade. Those technologies would capture carbon dioxide emitted by power plants fueled by either coal or natural gas and store it underground in geologic formations, such as deep saline formations, oil and gas fields, and coal beds that cannot be mined economically. For coal power plants with CCS, the coal would first be gasified and then the resulting gas ignited. Such integrated-gasification combined-cycle (IGCC) technology, which is already in use at a few power plants that do not capture carbon dioxide, allows for capturing it before combustion, when it is more concentrated.[2] Natural gas power plants with CCS use the same combined-cycle process as conventional natural gas plants but filter the carbon dioxide from the natural gas before combustion.[3]

Technologies Not Included in the Congressional Budget Office's Analysis

Oil and renewable energy technologies are not expected to compete with nuclear technology as a source of new base-load capacity nationwide. Because of high fuel costs, oil-fired generators are commercially competitive only in a few areas that have limited access to coal and natural gas, most notably, Hawaii. Intermittent technologies such as wind and solar power, which cannot operate much of the day, are not a source of base-load capacity. Other renewable technologies, like geothermal and hydroelectric power, can provide a more consistent flow of electricity but use resources that are mostly available in the West. Biomass technology can generate base-load electricity in certain parts of the country but is typically limited to small applications because fuel costs become prohibitive at large facilities.

1 The incentives provided by the Energy Policy Act of 2005 that are addressed in this study promote the research, design, and deployment of third-generation reactors. Additional incentives under that law promote the research and development of designs for a fourth generation, but those technologies are not expected to be deployable until around 2030.

2 In its analysis, the Congressional Budget Office also estimated the cost of electricity from new IGCC plants without CCS. Those results are not presented because they are roughly similar to the results for pulverized coal power plants.

3 For a more detailed discussion of technologies to capture and store carbon dioxide, see Congressional Budget Office, *The Potential for Carbon Sequestration in the United States* (September 2007), pp. 8–17.

Investors may have preferred new natural gas power plants as a source of additional base-load and peak capac ity because "combined-cycle" technology became an affordable and efficient technique for generating electricity.[6] As recently as 2000, EIA forecast that combined-cycle natural gas technology would be the least expensive means of generating base-load electricity until at least 2020. The expansion of natural gas capacity since 1994, rather than nuclear or coal capacity, could indicate that investors had similar expectations. However, EIA now projects that some of the recent increase in the price of natural gas will persist. (Figure 1-2 charts actual and projected natural gas prices.) If that assessment proves to be the case, natural gas capacity could be a relatively expensive source of base-load power.

Over the past few years, most likely in response to both the prospect of carbon dioxide charges and the incentives offered in EPAct, several utilities have begun planning new nuclear projects, which may signal the end of a 30-year hiatus in financing the construction of nuclear power plants. As of 2007, over a dozen utilities had announced their intention to file construction and operating licenses (COLs), which are obtained through the Nuclear Regulatory Commission (NRC), for roughly 30 nuclear plants.[7] Those plants would provide about 40,000 megawatts of new capacity.[8] For perspective, the roughly 100,000 megawatts of existing nuclear capacity currently provides about 20 percent of all electrical power in the United States.

Although the announcements reflect renewed interest in building new nuclear power plants, they do not indicate how much capacity utilities will ultimately build. Completing the revised design and licensing process is expected to cost about $100 million per plant, less than 5 percent of the anticipated cost for constructing a nuclear plant. Filing a COL application by the end of 2008 may be necessary for those projects to remain eligible for a share of the $7.5 billion (in nominal dollars) in production tax credits, but filing does not obligate an applicant to build the proposed plant.

Considerations Underlying Future Investment in Power Plants

New base-load capacity requires years to plan and build—roughly a decade in the case of nuclear technology. Because power plants can operate for many years (numerous power stations built in the first half of the previous century are still in use), new capacity is expected to replace existing capacity slowly in the absence of a cost advantage.

Utilities typically invest in new electricity-generating capacity to meet increases in demand or to replace facilities that have become too expensive to operate. When planning new power plants, both traditional regulated utilities (whose return on investment is largely determined by public utility commissions) and merchant generators (whose return is dictated by market outcomes) consider which technology generates electricity at the lowest cost. The cost of electricity from new capacity depends on the cost of building a plant, the cost of financing that construction, and the recurring cost of operating the plant (including the cost of fuel).

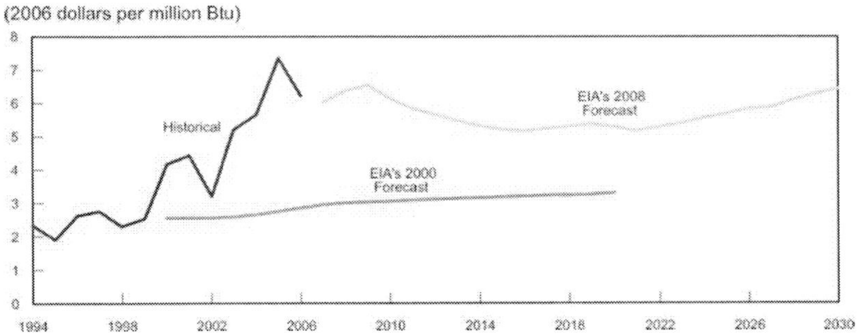

Source: Congressional Budget Office based on data from the Energy Information Administration (EIA).

Notes: Btu = British thermal unit.

Prices are expressed in real (inflation-adjusted) dollars using the gross domestic product price index; they represent average prices received by natural gas producers (in the lower 48 states) and, therefore, do not include the cost of delivering natural gas from the wellhead to the power plant. EIA provides the average price of natural gas delivered to power plants for the years from 1997 to 2006. On average, those prices exceed the reported prices by $0.70 per million Btu.

Figure 1-2. Actual and Projected Natural Gas Prices, 1994 to 2030

For the purposes of CBO's analysis, conventional coal plants are assumed to use pulverized coal technology, which produces energy by burning a crushed form of solid coal, and conventional natural gas plants are assumed to convert gas into electricity using combined-cycle turbines. The advanced nuclear technology considered in the analysis is confined to reactor designs that would be available in the next decade. At the same time, development continues on innovative fossil-fuel technologies that are designed to capture and store almost all of the carbon dioxide emitted while generating electricity. Over the next decade, investors will also consider those carbon capture-and-storage, or CCS, technologies as an alternative for new generation capacity.[9]

To compare the cost of alternatives for new generating capacity, CBO developed a reference scenario to serve as a benchmark against which the effects of both existing and prospective policy initiatives and future market conditions could be measured. Specifically, the reference scenario excludes the possible effects of carbon dioxide charges and EPAct incentives. On the basis of the assumptions underlying that scenario, CBO estimated the levelized costs of five technological alternatives. Levelized cost, a construct frequently used in analyzing investment in electricity generation, is the minimum price of electricity at which a technology generates enough revenue to pay all of the utilities' costs, including a sufficient return to investors.[10] Federal, state, and local policies can change the costs incurred by utilities by providing incentives, which shift costs or financial risk to the public, or by levying taxes on the utilities, which increases their costs.

Among federal laws and programs that influence investors' decisions about which technology to choose for new electricity-generating capacity are standard corporate tax laws, programs that support specific technologies by altering tax laws or providing other incentives, and taxes on specific goods whose production or consumption affects others. Carbon dioxide charges in particular could significantly increase the utilities' costs of generating electricity, which could reduce the amount of investment in new capacity if customers reduced their usage in reaction to higher prices. This analysis, however, focuses on how potential carbon dioxide charges and the provisions of EPAct would change the relative cost of alternatives for generating electricity and does not address the total quantity of base-load capacity that might be built once investors accounted for future demand. However, there is general agreement that demand will continue to grow with the population; accordingly, investing in additional capacity will be commercially viable unless the costs of supplying electricity rise significantly.

How Might Carbon Dioxide Charges Affect the Prospects of Investment in New Nuclear Plants?

Measuring the utilities' costs across a range of potential carbon dioxide charges indicates which technologies might be competitive, given certain assumptions about future legislative action and market outcomes. In general, the higher the costs to utilities of emitting carbon dioxide, the more competitive nuclear power would be because it is the only zero-emissions base-load technology.

In the absence of both emission charges and EPAct incentives, conventional fossil-fuel technology would dominate nuclear technology. But, even without EPAct incentives, if lawmakers enacted legislation that resulted in a carbon dioxide charge of about $45 per metric ton, nuclear generation would most likely become a more attractive investment for new capacity than conventional fossil-fuel generation (see the left panel of Figure 1-3). If the cost of emitting carbon dioxide was between $20 and $45 per metric ton, nuclear generation as an option for new capacity would probably be preferred over coal but not natural gas.

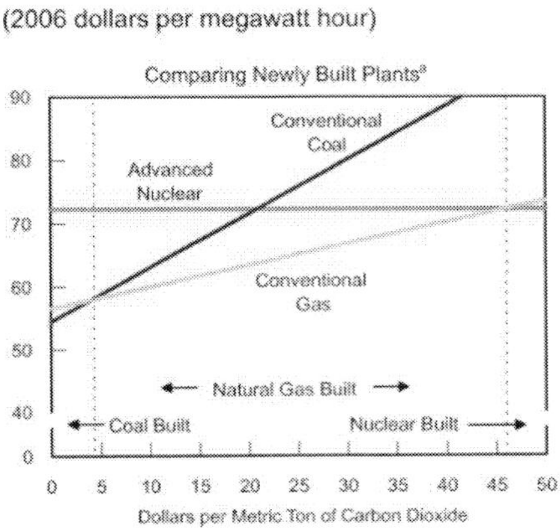

Source: Congressional Budget Office.

Notes: Electricity-generating capacity is measured in megawatts; the electrical power generated by that capacity is measured in megawatt hours. During a full hour of operation, 1 megawatt of capacity produces 1 megawatt hour of electricity, which can power roughly 800 average households.

Advanced nuclear technology refers to third-generation reactors. Conventional coal power plants are assumed to use pulverized coal technology, which produces energy by burning a crushed form of solid coal. Conventional natural gas power plants are assumed to convert gas into electricity using combined-cycle turbines.

These comparisons exclude the impacts of incentives provided under the Energy Policy Act of 2005.

a The levelized cost of new capacity using fossil-fuel technologies is not included in the figure, but those technologies generally have a higher levelized cost than nuclear technology.

b The levelized cost of new capacity using conventional natural gas technology is not included in the figure, but electricity produced using new natural gas capacity would be cheaper than continuing to operate existing coal capacity at carbon dioxide charges above $45 per metric ton.

Figure 1-3. Levelized Cost of Electricity under Carbon Dioxide Charges

Most of CBO's analysis focuses on technology choices for new power plants, but electricity from new capacity would still compete in the same market as that from existing capacity and could eventually begin to displace that capacity if the price of emitting carbon dioxide was sufficiently high. For instance, because utilities have already incurred construction costs for existing facilities, existing coal-fired power plants could be a less expensive source of electricity than new power plants under the range of carbon dioxide charges considered (see the right panel of Figure 1-3). If building new nuclear power plants proved to be less expensive than building new coal-fired plants but more expensive than using existing coal capacity, additions to nuclear capacity would be limited to meeting increases in electricity usage. Despite the high carbon intensity of conventional coal technology, continuing to operate existing coal-fired plants would remain a relatively inexpensive source of electricity until carbon dioxide charges reached about $45 per metric ton.

How Does the Energy Policy Act of 2005 Affect the Prospects of Investment in New Nuclear Plants?

During the next several years, the incentives put in place or extended by EPAct could significantly improve the relative cost of at least the first few nuclear plants built. (See Table 1-1 for an overview of EPAct incentives.) In contrast to carbon dioxide charges, which make nuclear alternatives attractive by increasing the cost of fossil-fuel alternatives, subsidies for new nuclear projects directly reduce the cost of nuclear plants in comparison to fossil-fuel options.

The largest incentives available under EPAct are a production tax credit and a loan guarantee program. The tax credit provides up to $18 in tax relief per megawatt hour of electricity produced at qualifying power plants during the first eight years of operation. (For comparison, the average wholesale price of electricity in 2005 was about $50 per megawatt, on average.)[11] The loan guarantee program provides a federal guarantee on debt that covers as much as 80 percent of construction costs. (Debt for building capacity has been financed at roughly 1 percentage point to 5 percentage points above the Treasury-bill rate.)[12] The loan guarantee program also applies to utilities employing innovative fossil-fuel or renewable technologies.

Without the incentives offered to investors who chose nuclear and other innovative generating technologies, conventional coal technology would be the least expensive means of generating electricity (see Figure 1-4). Generating electricity with nuclear technology would be roughly 35 percent more expensive than using conventional coal technology and 30 percent more expensive than using natural gas capacity. Accordingly, investment in nuclear capacity would be unlikely in the absence of carbon dioxide charges and EPAct incentives.

Table 1-1. Incentives Provided by the Energy Policy Act of 2005

Program	Technology	Incentives	Federal Cost
Research and Development			
Nuclear Power 2010	Advanced nuclear	DOE covers one-half of FOAK costs for licensing and design	$281 million (nominal dollars through 2007)
FutureGen	Innovative coal	DOE shares the cost of developing new facilities	$97 million (nominal dollars through 2007)
Investment			
Loan Guarantee	Eligible technologies (including advanced nuclear and innovative coal)	Provides low-cost debt financing on up to 80 percent of construction costs (the Treasury reimburses the lender in cases of default)	The utility pays the Treasury for administrative and subsidy costs, but tax revenues diminish under debt financing[a,b]
Delay Insurance[c]	Advanced nuclear	Applies to the first six nuclear plants covered by DOE. Compensates for certain delays in operation providing up to $500 million apiece for the first two plants and $250 million each for the next four	Utility pays subsidy cost[a]
Investment Tax Credit	Innovative coal[d]	Provides tax credits for up to 20 percent of a project's construction costs	Less than the dollar value of the credits[e]
Production			
Production Tax Credit	Advanced nuclear	Up to $18/MWh over the first eight years of operation for new nuclear plants [f, g]	A reduction in tax revenues up to the value of the credits issued[h]
Limited Liability[i]	Nuclear	Applies to plants built through 2025. The nuclear industry would not be responsible for damage exceeding $10.6 billion from a nuclear accident [j]	Probably small in terms of expected costs (see Box 3-1 in Chapter 3)
Tax Treatment of Decommissioning Funds	Nuclear	Extends to plants owned by merchant generators. Funds taxed at a reduced rate (20 percent)[k]	A reduction in tax revenues

Source: Congressional Budget Office.

Notes: CCS=carbon capture and storage; DOE= Department of Energy; EPAct= Energy Policy Act of 2005; FOAK=first-of-a-kind costs; MW= megawatt; MWh=megawatt hour.

Electricity-generating capacity is measured in MW; the electrical power generated by that capacity is measured in MWh. During a full hour of operation, 1 MW of capacity produces 1 MWh of electricity, which can power roughly 800 average households.

Advanced nuclear technology refers to third-generation reactors; nuclear technology not designated "advanced" may use first- and second-generation reactors. Innovative coal and natural gas technologies are assumed to capture and store most carbon dioxide emissions.

a The subsidy cost is the cost the government is expected to incur by guaranteeing debt or insuring against certain delays. In the absence of appropriations for that purpose, the utility must pay the cost; to date, the Congress has not made such appropriations.

b By making debt financing cheaper, the loan guarantee program increases the amount of financing for which interest payments are tax deductible.

c Delay insurance is authorized by EPAct under the title "Standby Support for Certain Nuclear Plant Delays."

d Credits have been available for both innovative coal technology and conventional (pulverized) coal technology that meet specific efficiency and environmental criteria. This study considers only the investment tax credit for innovative coal technology because the credits for conventional coal technology have been awarded already. An investment tax credit is also provided for new solar capacity.

e The net reduction in tax payments for an investment tax credit is less because applying the credit reduces the amount of capital cost that may be deducted through standard corporate tax law. For the percentage of capital cost on which a credit is claimed, 50 percent of the standard tax deduction may be taken.

f Production tax credits are also available for technologies that use renewable energy sources.

g The production tax credit for advanced nuclear technology is not adjusted for inflation; consequently, the maximum value of the credit is likely to decrease substantially by the time advanced nuclear power plants begin operating. In addition, if more than 6,000 MW of capacity qualified, the credit would then be divided proportionally among all qualified capacity. For example, if 12,000 MW of capacity qualified, then the owners of each plant would receive a maximum credit of $9/MWh. The per-MWh value of the credit would also be reduced if the nuclear power plant operated at 80 percent above capacity.

h The value of credits used cannot exceed the utility's alternative minimum tax liability.

i Limited liability is provided by the Price-Anderson Nuclear Industries Indemnity Act, which was extended under EPAct.

j The $10.6 billion figure assumes that the 104 operating commercial reactors will remain operational. The industry's liability per accident equals roughly $300 million plus $96 million multiplied by the number of operating commercial reactors.

k A utility may deduct the cost of decommissioning as payments are made to the funds, although deductions typically are not made until the service (decommissioning) is performed.

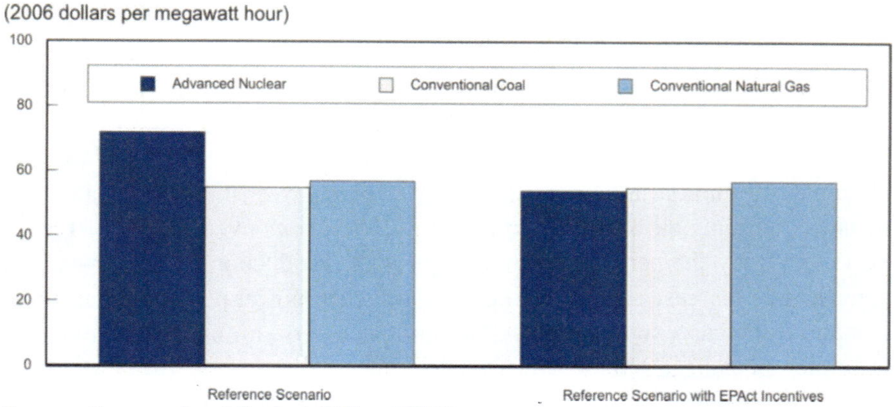

Source: Congressional Budget Office (CBO).

Notes: EPAct= Energy Policy Act of 2005.

CBO's reference scenario excludes both the effects of prospective carbon dioxide constraints and the impact of EPAct incentives.

Electricity-generating capacity is measured in megawatts; the electrical power generated by that capacity is measured in megawatt hours. During a full hour of operation, 1 megawatt of capacity produces 1 megawatt hour of electricity, which can power roughly 800 average households.

The estimate of the effect of EPAct incentives assumes that advanced nuclear technology receives the maximum production tax credits and loan guarantees. The production tax credits are shared among 6,000 megawatts or less of advanced nuclear capacity, and loan guarantees cover 80 percent of construction costs.

Advanced nuclear technology refers to third-generation reactors. Conventional coal power plants are assumed to use pulverized coal technology, which produces energy by burning a crushed form of solid coal. Conventional natural gas power plants are assumed to convert gas into electricity using combined-cycle turbines.

Figure 1-4. Levelized Cost of Electricity With and Without EPAct Incentives

Nuclear power would be a competitive technology for a few new power plants, however, if those plants received the maximum benefits that could be provided under EPAct. Most of the reductions in the cost for those plants would come from the production tax credit and loan guarantees. Other incentives—such as preferential tax treatment for decommissioning funds and limited liability protection—would have a relatively small effect on the cost of nuclear capacity.[13] Incentives covering first-of-akind (FOAK) costs could be crucial for attracting financing for the first nuclear plants of each advanced reactor design, although those incentives might not directly affect the cost of subsequent plants. The investment tax credit and loan guarantees for innovative coal plants with CCS and loan guarantees for innovative natural gas power plants with CCS reduce the utilities' costs for those technologies but not by enough to make them less expensive than nuclear power plants that qualify for EPAct incentives or conventional fossil-fuel power plants.

The cost of new nuclear capacity would probably be higher if utilities attempted to build a large number of power plants over the next decade. For instance, building all of the 30 proposed nuclear plants over the next 10 to 15 years—roughly the period of availability for the production tax credit—could significantly increase construction costs for nuclear power plants by increasing demand for scarce components that are necessary to build reactors (for example, specialized steel forgings).

A large wave of additions could also lead to higher costs by reducing the value of the production tax credits or by exhausting coverage under the loan guarantee program. EPAct limits production tax credits for nuclear power plants to a total of $7.5 billion, which means that each eligible plant's allotment of credits would decrease if more than 6,000 megawatts of capacity (roughly the capacity of five plants) qualified for the credit.[14] CBO's analysis incorporates the assumption that no more than 6,000 megawatts of capacity would qualify. Thus, the comparison of costs is intended to indicate only whether nuclear technology would be a commercially viable choice for up to a few nuclear power plants. For gauging the long-run competitiveness of nuclear generation, potential carbon dioxide charges are more likely to influence the development of new nuclear capacity than EPAct incentives.

Uncertainties Posed by Future Market Conditions and the Possibility of Carbon Dioxide Constraints

The commercial viability of nuclear capacity depends both on generators' perceptions of future market conditions at the point they consider committing to the construction of a plant—which might not occur for a few years—and the return that investors would require if confronted with carbon dioxide charges. An array of factors—recent volatility in natural gas prices and construction costs, nuclear power's history of construction cost overruns, and uncertainty about future policy on carbon dioxide emissions—suggests that a wide range of costs are plausible for each of the base-load technologies. Those ranges in costs for new power plants demonstrate that each technology faces considerable uncertainty.

Costs under Alternative Market Conditions

The assumptions used in the reference scenario are intended to represent investors' perceptions, but even if those base-case assumptions accurately portray the current outlook, unanticipated events may alter those perceptions before investors make binding commitments to nuclear capacity. CBO compared levelized costs under several plausible future scenarios for fuel, construction, and financing costs to identify which technology utilities would probably choose for new capacity under a broad range of circumstances.

Cost of Fuel

In the reference scenario, CBO assumed that natural gas prices in the future would be similar to average prices observed since 2000. However, if natural gas prices fell to levels seen in the 1990s and carbon dioxide emissions remained unconstrained, conventional natural gas technology would be the cheapest source of base-load capacity, even if nuclear technology benefited from EPAct incentives. Conversely, if natural gas prices continued to rise, as they have since the 1990s, natural gas technology would be unlikely to be a competitive alternative for base-load electricity generation. (Table 1-2 shows the cost of generating electricity using each technology under the reference scenario and under alternative market conditions that involve substantial deviations from the base-case assumptions about costs for fuel and construction.)

Although uranium prices have fluctuated widely over the past few years, fuel costs have historically accounted for a small share of the cost of nuclear capacity. EIA projects that the price of nuclear fuel will increase by about 40 percent over the next decade—a trend that CBO incorporates in its base-case assumptions—but if nuclear fuel prices stay at the relatively low 2006 level, the overall cost of nuclear technology would decrease by only 3 percent. (Table 1-2 shows that the cost of nuclear technology is largely insensitive to changes in fuel prices.) Utilities investing in new nuclear power plants would incur most of the cost of that technology during construction.

Cost of Construction

Historically, construction costs for nuclear facilities have been roughly double initial estimates. NRC's revised licensing process for nuclear power plants is expected to reduce midconstruction modifications, which were blamed for many cost overruns in the past. Moreover, vendors argue that advanced reactors will have lower construction costs because they have fewer parts than older reactors. As a result, CBO's base-case assumption for construction costs is about 25 percent lower than the historical average—a figure that reflects recent experience in the construction of advanced reactors in Japan. If those factors turned out not to reduce construction costs in the United States, nuclear capacity would probably be an unattractive investment even with EPAct incentives, unless substantial carbon dioxide charges were imposed.

Cost of Financing

The cost of financing construction is substantial for all technologies but particularly so for capital-intense technologies. In CBO's base-case assumptions, the cost incurred to finance commercially viable projects did not depend on which technology was used for a given project. That assumption would be justified if volatility in natural gas prices and the prospect of constraints on carbon dioxide emissions created cost uncertainties for conventional fossil-fuel technologies that were similar in magnitude to the uncertainties facing investment in nuclear technology.[15]

Costs under Carbon Dioxide Constraints

Utilities that invest in coal capacity face substantial cost uncertainties because of the prospect of future carbon dioxide constraints and the unknown stringency of such constraints. (The last two columns of Table 1-2 show the cost for new power plants by technology associated with two proposed carbon dioxide constraints.) A policy that constrained carbon dioxide emissions to 2008 levels is projected to increase the cost of electricity from conventional coal-fired plants by roughly one-half and a more stringent constraint would more than double the cost.[16] The projected cost of electricity from conventional natural gas capacity is about half as sensitive to such constraints. Because nuclear plants do not emit carbon dioxide, their levelized costs would be unaffected by the stringency of carbon charges.

Table 1-2. Levelized Cost of Electricity under Alternative Market and Policy Conditions

(2006 dollars per megawatt hour)

	Variations in Future Market Conditions					Variations in Future Carbon Dioxide Policy	
	Reference	Fuel Costs		Construction Costs		Emissions Capped at 2008 Level	Emissions Capped at About 85% below 2008 Level by 2050
	Scenario	High (+100%)	Low (-50%)	High (+100%)	Low (-50%)		
Advanced Nuclear	72	80	68	121	48	72	72
Conventional Coal	55	70	47	83	40	80	128
Conventional Natural Gas	57	97	36	69	51	67	86

Source: Congressional Budget Office.

Notes: Electricity-generating capacity is measured in megawatts; the electrical power generated by that capacity is measured in megawatt hours. During a full hour of operation, 1 megawatt of capacity produces 1 megawatt hour of electricity, which can power roughly 800 average households.

Advanced nuclear technology refers to third-generation reactors. Conventional coal power plants are assumed to use pulverized coal technology, which produces energy by burning a crushed form of solid coal. Conventional natural gas power plants are assumed to convert gas into electricity using combined-cycle turbines.

Alternative market and policy conditions are overlaid on the reference scenario, which excludes both the effects of prospective carbon dioxide constraints and the impact of incentives provided under the Energy Policy Act of 2005.

2. Framing the Analysis: Base-Case Assumptions and the Effects of Policy

To assess the competitiveness of advanced nuclear technology in comparison with other base-load options, the Congressional Budget Office estimated the levelized cost of alternatives under a reference scenario reflecting the agency's best judgment about future market conditions and the policy environment before the enactment of the Energy Policy Act of 2005, and under alternatives that consider the effects of both carbon dioxide charges and EPAct incentives. To calculate those costs, CBO adopted base-case assumptions about an array of technical and economic choices confronting investors in new electricity-generating capacity.[17]

Some of the assumptions underlying this analysis are very uncertain. For instance, large power plants using carbon capture-and-storage technologies exist only as blueprints, which makes predicting their costs difficult. Also subject to uncertainty and controversy are the costs of building a new nuclear plant in the United States, where no reactor has been ordered since the 1 970s, when substantial cost overruns were the rule. Even the risks of investing in conventional coal power plants are heightened because utilities cannot anticipate with certainty whether charges on carbon dioxide emissions will be imposed in the future and, if so, at what level. For those and other reasons, financial markets may require a higher rate of return for any investment in new base-load capacity. To account for those uncertainties, CBO assessed the levelized cost of alternatives over a range of values for critical assumptions about plant costs, fuel costs, and the rates of return required by investors to finance new capacity. The results of such calculations indicate the potentially significant impact of those uncertainties on CBO's estimates of the levelized costs of the various technologies.

Levelized Cost Analysis

Levelized cost is the minimum price at which a technology option produces electricity and generates enough revenue to pay all of a utility's costs and still provide a sufficient return to investors. In its analysis, CBO projected expected cash flows in order to find the minimum real (inflation-adjusted) price of electricity at which revenue exceeded costs by enough to encourage investment in the construction of new capacity based on each technology. (A detailed description of the approach used to estimate levelized costs in this study can be found in CBO's Web supplement, *The Methodology Behind the Levelized Cost Analysis.*)

Levelized costs affect investment decisions made by both merchant generators and regulated utilities. If the levelized cost of a technology exceeded anticipated prices for electricity, merchant generators would be unlikely to invest in new capacity based on that technology because the expected return would not justify the amount of risk they would have to incur. State utility commissions commonly direct regulated utilities to meet anticipated demand for new capacity using the technology with the lowest levelized cost.

Both types of utilities typically fund costs that are incurred before a plant begins operating through a combination of debt and equity financing. The revenue that results from the sale of electrical power is first used to pay the plant's operating costs, including the purchase of fuel. After deductions are made for corporate income taxes and debt payments, the remaining revenue is paid to equity-holders. CBO estimated the lowest constant real price of electricity at which the return to equity was adequate to attract the investment for up-front costs.

Whereas utilities' decisions are made on a site-by-site basis, the levelized costs estimated by CBO are intended to give a representative cost for each technology. One technology would probably not have the lowest cost in all parts of the country and other factors could be considered, but a technology with the lowest representative levelized cost would most likely be a common choice for new capacity.

Another consideration is that levelized costs do not indicate which technology produces electricity most efficiently because they capture the utility's cost of generating electricity rather than the actual cost. Government incentives that directly subsidize electricity generation or transfer financial risk from investors to the public decrease levelized costs in comparison to actual costs. Because corporate income taxes increase utilities' costs relative to actual costs, they could make efficient, capital- intense technologies relatively costly. Last, levelized costs include only those costs that markets and current laws require utilities to pay. For example, a carbon-emitting technology might have the lowest levelized cost in the absence of a carbon dioxide charge but not be the most efficient technology because the levelized cost would not account for the damage caused by carbon dioxide emissions.

Base-Case Assumptions

The base-case assumptions necessary to estimate the levelized cost of plants that employ alternative technologies include the cost of building a plant as if it was built and paid for immediately (so-called overnight costs); the return that investors require to finance that construction and subsequent operations; and the cost of operating the plant (largely composed of fuel costs). As a first approximation, CBO relied on the Energy Information Administration's most recent projections for its base-case assumptions and compared those with assumptions adopted in two prominent studies of generation alternatives, one conducted by researchers at the Massachusetts Institute of Technology (MIT) and the other by analysts at the International Energy Agency.[18]

Construction Costs

CBO's base-case assumptions include overnight costs of about $2.4 million for each megawatt of capacity for new nuclear plants and innovative coal plants but lower costs for conventional coal, conventional natural gas, and innovative natural gas technologies. For nuclear and innovative coal and natural gas technologies, the assumptions are intended to represent plants built over the next decade but do not incorporate the first-of-a-kind costs that are assumed to be covered by federal research and development programs. The estimate for nuclear plants, taken from the EIA's most recent analysis, is roughly 10 percent above the estimate of overnight costs used in MIT's study, which was published in 2003, before construction costs for most types of power plants surged. CBO also calculated construction costs for each technology using alternative assumptions designed to capture plausible variations in those costs. For nuclear and innovative coal technologies, CBO considered construction costs ranging from about $1.2 million per megawatt of capacity to roughly $4.8 million per megawatt of capacity. The breadth of that range reflects the uncertainty associated with the cost of building new nuclear plants in the United States and is wide enough to capture plausible further increases in construction costs, which could affect conventional fossil-fuel plants as well.

CBO's assumption about the cost of building new nuclear power plants in the United States is particularly uncertain because of the industry's history of construction cost overruns. For the 75 nuclear power plants built in the United States between 1966 and 1986, the average actual cost of construction exceeded the initial estimates by over 200 percent (see Table 2-1). Although no new nuclear power plants were proposed after the partial core meltdown at Three Mile Island in 1979, utilities attempted to complete more than 40 nuclear power projects already under way. For those plants, construction cost overruns exceeded 250 percent.[19] (An average of 12 years elapsed between the start of construction and the point at which the plants began commercial operation. The overruns in overnight costs did not include additional financing costs that were attributable to post- accident construction delays.)[20]

Table 2-1. Projected and Actual Construction Costs for Nuclear Power Plants

Construction Starts		Average Overnight Costs[a]		
Year Initiated	Number of Plants[b]	Utilities' Projections (Thousands of dollars per MW)	Actual (Thousands of dollars per MW)	Overrun (Percent)
1966 to 1967	11	612	1,279	109
1968 to 1969	26	741	2,180	194
1970 to 1971	12	829	2,889	248
1972 to 1973	7	1,220	3,882	218
1974 to 1975	14	1,263	4,817	281
1976 to 1977	5	1,630	4,377	169
Overall Average	**13**	**938**	**2,959**	**207**

Source: Congressional Budget Office (CBO) based on data from Energy Information Administration, *An Analysis of Nuclear Power Plant Construction Costs,* Technical Report DOE/EIA-0485 (January 1, 1986).

Notes: Electricity-generating capacity is measured in megawatts (MW); the electrical power generated by that capacity is measured in megawatt hours (MWh). During a full hour of operation, 1 MW of capacity produces 1 MWh of electricity, which can power roughly 800 average households.

The data underlying CBO's analysis include only plants on which construction was begun after 1965 and completed by 1986.

Data are expressed in 1982 dollars and adjusted to 2006 dollars using the Bureau of Economic Analysis's price index for private fixed investment in electricity-generating structures. Averages are weighted by the number of plants.

a. Overnight construction costs do not include financing charges.

b. In this study, a nuclear power plant is defined as having one reactor. (For example, if a utility built two reactors at the same site, that configuration would be considered two additional power plants.)

The base-case assumption adopted in this analysis for nuclear power plants' overnight costs recognizes that history but also allows for countervailing factors, such as changes in the U.S. regulatory process and other countries' recent experience with new reactor designs. In 1989, the Nuclear Regulatory Commission developed an alternative process for obtaining the licensing necessary to operate a nuclear power plant. That revised process is intended to reduce cost uncertainties by allowing utilities to fulfill more regulatory requirements before beginning construction, thereby reducing midconstruction design changes that contributed to overruns in the past.

The experience of a Japanese utility, the Tokyo Electric Power Company (TEPCO), in the mid-1990s also appears to support CBO's base-case assumption about construction costs. According to the 2003 MIT study, verifiable data indicate that TEPCO constructed two advanced boiling-water reactors at costs and schedules close to manufacturers' estimates.[21] However, a Finnish utility that is building a reactor based on a different design, an advanced pressurized-water reactor, continues to have difficulty adhering to original cost estimates. By 2007, that project, initially estimated to cost 3 billion, had fallen at least 18 months behind schedule, causing costs to increase by 700 million.[22]

Table 2-2. Financial Risk Assumptions in Comparable Studies

Study	Real Rate of Return (Percent)	Capital Recovery Period[a] (Years)
CBO	8-^{3}/4 – 12-^{1}/2	40
EIA	9-^{1}/4	20
IEA	8 – 11	40 – 25
MIT	8 – 11-^{1}/4	25 – 40

Source: Congressional Budget Office (CBO).

Notes: EIA = Energy Information Administration; IEA = International Energy Agency; MIT= Massachusetts Institute of Technology.

Real rates of return are rounded to the nearest quarter of a percentage point. CBO calculated those rates on the basis of the inflation rates and nominal rates of return used in each study. The underlying nominal rates of return for the studies conducted by EIA and IEA represent the weighted average cost of capital assumed in those studies. The nominal rates of return for the MIT study were constructed by taking the ratio of financing charges to balances in CBO's replication of the MIT model.

a. The capital recovery period represents the number of years over which revenue from the sale of electricity is used to repay debt or equityholders.

Financing Costs

Even if construction proceeds on schedule, utilities still incur substantial financing costs because power plants take years to build, and financing costs for construction extend over the decades that a plant generates electricity. CBO's assumptions about financing costs are a synthesis of the financial analyses presented in the studies by EIA and MIT. Those assumptions are encapsulated by the real rate of return that investors require to assume the risk of paying up-front construction costs.[23] CBO used a real rate of return of 10 percent, which falls within the range of rates of return given in the other studies (see Table 2-2).[24] The 10 percent rate of return was used for each technology, reflecting that the level of financial risk is similar across commercially viable projects. The MIT study assumed that a higher rate of return would be required for nuclear technology than for conventional fossil-fuel technologies; however, that 2003 study was published before much of the volatility in natural gas prices and when future federal carbon dioxide constraints may have appeared less likely.[25] But nuclear plants could still be a riskier investment than competing alternatives, or the rate of return could vary for all technologies. In addition to the base-case assumption of a 10 percent rate of return, CBO considered the competitiveness of nuclear power under lower and higher rates (as shown in Table 2-2).

Fuel Costs

The cost of fuel is one of the most significant operating costs included in CBO's estimates of the levelized cost of options for generating electricity. The base-case assumption for nuclear power is that $8 (in 2006 dollars) in fuel costs are incurred for each megawatt hour of electricity generated (see Table 2-3). That contrasts with $16 for conventional coal-fired plants and $40 for conventional natural gas plants.[26] Those assumptions are based on long-term projections by EIA. In the past, fuel costs have proved difficult to predict, particularly the price of natural gas. (See Figure 2-1 for fluctuations in fuel prices between 1995 and 2006.) In addition to those base-case assumptions, CBO also estimated levelized costs using alternative assumptions intended to capture most plausible variations in fuel costs.

The cost of disposing of the used (spent) fuel generated by nuclear fission is currently unique to that fuel source. (However, if carbon dioxide charges were imposed in the future, they could be seen as similar because utilities could have to assume the full cost of their fuel choice— that is, they would have to pay for the damage inflicted by emitting carbon dioxide.) CBO's levelized cost estimate for nuclear power includes a $1 per megawatt hour charge to cover the cost of such disposal.[27]

Additional Assumptions

Other important considerations in CBO's analysis include the fixed cost of operating and maintaining a power plant (that is, the costs that do not vary with the amount of electricity produced) and the percentage of maximum electricity production achieved by the plant (the "capacity factor"). For nuclear plants, fixed operating and maintenance expenses include the cost of providing security and monitoring safety systems. On the basis of EIA's analysis, the combination of such costs is assumed to be five times those incurred by a conventional natural gas coal plant and twice those incurred by a conventional coal plant. The fixed operating costs of innovative coal and natural gas plants are assumed to be higher than those of conventional fossil-fuel alternatives but lower than those of nuclear power plants.

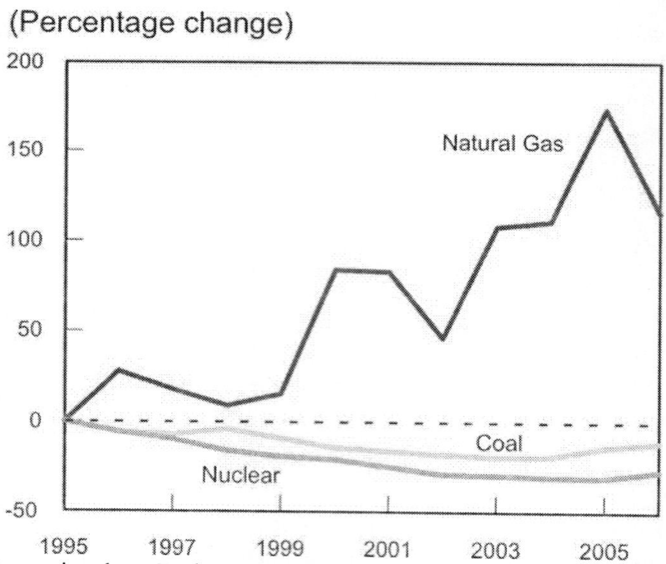

Source:Congressional Budget Office based on data from the Energy Information Administration (EIA).

Note: The percentage changes are based on prices in 2006 dollars, with adjustments for inflation made using the gross domestic product price index. Prices for all fuels equal the average cost at which those fuels are delivered to power plants, as measured by EIA.

Figure 2-1. Historical Volatility in Fuel Prices

Table 2-3. Key Assumptions underlying CBO's Reference Scenario

	Advanced Nuclear	Conventional Coal	Conventional Natural Gas	Innovative Coal	Innovative Natural Gas
Construction					
Time (Years)	6	4	3	4	3
Overnight costs (Thousands of dollars per MW)[a]	2,358	1,499	685	2,471	1,388
Operating Costs					
Fuel (Dollars per MWh)	8	16	40	17	52
Fixed operation and maintenance (Dollars per MWh)	8	4	1	6	3

Source: Congressional Budget Office based on data from the Energy Information Administration.

Notes: Electricity-generating capacity is measured in megawatts; the electrical power generated by that capacity is measured in megawatt hours. During a full hour of operation, 1 megawatt of capacity produces 1 megawatt hour of electricity, which can power roughly 800 average households.

The reference scenario excludes both the effects of prospective carbon dioxide charges and the impact of incentives provided under the Energy Policy Act of 2005.

Advanced nuclear technology refers to third-generation reactors. Conventional coal power plants are assumed to use pulverized coal technology, which produces energy by burning a crushed form of solid coal. Conventional natural gas power plants are assumed to convert gas into electricity using combined-cycle turbines.

Values are expressed in 2006 dollars.

a. Overnight construction costs do not include financing charges. Financing charges are addressed separately in Table 2-2.

Capacity factors range from about 80 percent to 90 percent, depending on the technology. Those factors represent the maximum rate at which each technology could physically operate, as determined by EIA. Historically, utilities have utilized natural gas capacity at much lower rates because, in comparison with coal-fired and nuclear technologies, natural gas has typically been an expensive source of base-load power. But because this study evaluates the competitiveness of natural gas as a base-load alternative, CBO assumed that options using that fuel would operate at their maximum capacity factor.

Accounting for the Effects of Policy

Expanding the reference scenario to include the effects of carbon dioxide charges and EPAct incentives changes CBO's estimates of the levelized cost—and potential competitiveness—of each technological option. Carbon dioxide charges would raise the levelized cost of fossil-fuel alternatives but not the cost of nuclear power. Conversely, EPAct incentives reduce the levelized cost of nuclear power and innovative fossil-fuel options in comparison with that of conventional fossil-fuel technologies.

In addition to the uncertainty inherent in the base-case assumptions, some of the assumptions linking policy to levelized cost are subject to uncertainty. In particular, when estimating levelized cost, CBO assumed that the full benefits of EPAct incentives would be available to both nuclear and innovative fossil-fuel options. In some instances, CBO also used simplifying assumptions to incorporate policy into levelized cost. For example, carbon dioxide charges were assumed to be levied only on smoke-stack emissions (those resulting directly from the operation of a power plant). In reality, all technologies have the potential to produce additional emissions from the construction and decommissioning of a facility, as well as from the processing of fuel. Such additional life- cycle emissions are not included in this analysis because they are difficult to measure precisely and are probably an order of magnitude smaller than the stack emissions from coal or natural gas power plants.[28]

Accounting for the Effects of Carbon Dioxide Charges

Carbon dioxide constraints could encourage the use of nuclear technology by increasing the cost of generating electricity with fossil fuels. The effect is most pronounced for coal, which emits nearly a metric ton of carbon dioxide for every megawatt hour of electricity produced. The effect on conventional generators fueled by natural gas would be less because they emit carbon dioxide at roughly half the rate of the average coal plant.

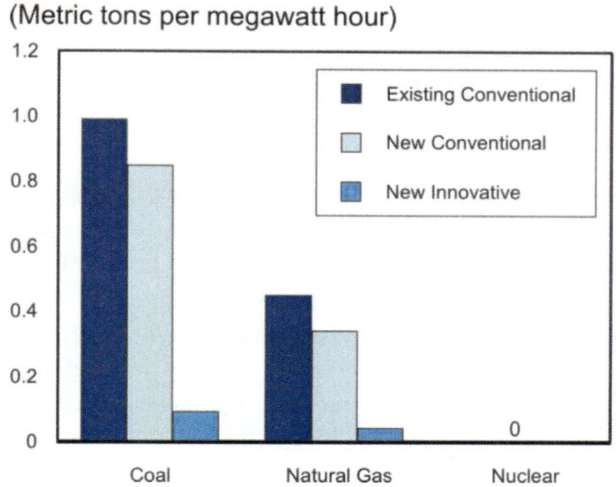

Source: Congressional Budget Office (CBO).

Note: Electricity-generating capacity is measured in megawatts (MW); the electrical power generated by that capacity is measured in megawatt hours (MWh). During a full hour of operation, 1 MW of capacity produces 1 MWh of electricity, which can power roughly 800 average households.

Conventional coal power plants are assumed to use pulverized coal technology, which produces energy by burning a crushed form of solid coal. Conventional natural gas power plants are assumed to convert gas into electricity using combined-cycle turbines. Both innovative coal and innovative natural gas technologies are assumed to capture and store most carbon dioxide (CO_2) emissions.

CBO's analysis assumes that coal contains approximately 0.095 metric tons per million British thermal units (Btu) of CO_2 and that natural gas contains 0.054 metric tons per million Btu. It also assumes that existing conventional coal technology burns 10.463 million Btu of coal per MWh of electricity and that existing conventional natural gas technology burns 8.401 million Btu of natural gas per MWh of electricity. See CBO's Web supplement for assumptions underlying the analysis.

Figure 2-2. Carbon Dioxide Emissions of Base-Load Technologies for Generating Electricity

Because competing base-load alternatives emit carbon dioxide, the attractiveness of financing a new nuclear power plant depends on investors' expectations about the costs of emitting carbon dioxide over the operating life of that plant. To the extent that carbon dioxide charges are expected, investment in new nuclear capacity would be more attractive relative to both the construction of new fossil-fuel capacity and the continued use of existing fossil-fuel capacity. Many investors appear to anticipate some form of carbon dioxide charge in the near future; a survey conducted by Cambridge Energy Research Associates in 2006 found that about 80 percent of utility executives expected a carbon dioxide charge to be implemented within the next 10 years.[29]

Although the imposition of carbon dioxide constraints would not directly decrease the cost of operating nuclear power plants, such a policy would increase the cost of operating fossil-fuel power plants, which in all their variants emit at least some carbon dioxide, and consequently make new nuclear capacity a more attractive source of base-load generation. Newly built power plants based on conventional fossil-fuel technology are designed to burn fuel more efficiently than plants built in the past, but their emissions would still be substantial enough for the cost of producing electricity to be sensitive to carbon dioxide charges. Innovative fossil-fuel power plants that capture and store carbon dioxide are assumed to emit only about 10 percent of the carbon dioxide discharged into the atmosphere by the lowest emitting conventional plants that burn fossil fuel—but they still emit carbon dioxide. As of 2007, such carbon dioxide capture-andstorage technologies had not been used at commercial power plants, but those technologies could be an option for new base-load capacity by the time new nuclear plants were deployed and might be the most competitive alternative to nuclear technology under carbon dioxide charges.[30]

CBO estimated the cost of emitting carbon dioxide using hypothetical charges based on the levels of carbon intensity for coal and natural gas reported in the MIT study. Those hypothetical charges were assumed to be proportional to the amount of carbon dioxide emitted by each technology (see Figure 2-2).

Accounting for the Effects of Energy Policy Act Incentives

EPAct provides or extends numerous incentives for generating electricity using nuclear, renewable, and innovative fossil-fuel technologies. In general, the incentives lower the cost of nuclear and other innovative technologies in comparison to conventional fossil-fuel alternatives. Among the programs reauthorized under EPAct were the Nuclear Power 2010 program and FutureGen (originally authorized in 2002 and 2003, respectively).[31] Under those programs, the federal government shares with industry the cost of researching, developing, and deploying advanced nuclear power plants and innovative coal-fired plants that incorporate CCS technology. Other incentives—in particular, loan guarantees, investment tax credits, and insurance against regulatory delays—are intended to encourage investment in nuclear power and innovative fossil-fuel technologies. Another set of EPAct policies provides production incentives for operating advanced nuclear power plants once construction is complete. (Table 1-1 on page 10 lists and describes in more detail the incentives provided by EPAct.)

Research and Development Incentives

The lessons learned in developing and implementing a particular design for a new power plant could reduce the cost of building additional power plants with similar designs. First-of-a-kind costs could be especially large in the nuclear industry because the technology is complex, and utilities have little experience in navigating the revised regulatory process for obtaining a construction and operating license. To the extent that the original utility does not have exclusive rights to build any additional plants, part of the gains in knowledge and experience arising from the initial investment could be captured by other utilities that build plants later. As a result, utilities might underinvest in new technologies because they would not retain enough of the benefit such investment produced.[32]

The Nuclear Power 2010 pays a share of FOAK costs for advanced nuclear technology—to increase the amount of investment in its development.[33] To estimate the reduction in costs attained by plants benefiting from that program, CBO included FOAK costs for licensing and design in supplementary analysis.

Investment Incentives

A second set of incentives encourages investment in the construction of power plants that use advanced nuclear and innovative fossil-fuel technologies. They include a loan guarantee program that insures the debt for such technologies, another insurance program that provides advanced nuclear technologies protection against the cost of certain delays in the start of operation, and tax incentives for investment in innovative coal technologies.

Such incentives could be viewed as countering negative effects on investment caused by taxes on capital income. For instance, corporate income taxes, as well as taxes on capital gains, dividends, and interest income, act as proportional surcharges on investing in the construction of power plants. Such taxes could cause utilities to prefer technologies that were less capital-intensive. In particular, capital costs make up a relatively large portion of the cost of producing electricity using nuclear or innovative coal power plants because those plants are relatively expensive to build. As a result, taxes on capital income might encourage utilities to build conventional fossil-fuel power plants, which have lower capital costs. Investment incentives could counter potential bias against capital-intensive technologies caused by taxes on capital income.

Loan guarantees and insurance against delays reduce the financial risk of investing in advanced nuclear power plants by transferring risk to the public. The reduced risk means investors would incur lower costs for financing construction and other activities before a plant began operating. However, economic theory suggests that such incentives cause recipients to invest in excessively risky projects because they do not bear all the cost of a project's failure. The federal government also provides investment subsidies through investment tax credits, which reduce tax liability in proportion to construction expenditures.

Production Incentives

A third set of incentives encourages not only the construction of nuclear plants but also their continued operation. Those incentives indirectly encourage investment by making operation more profitable. Incentives supporting operation include a production tax credit, a limit on liability for nuclear accidents, and a tax incentive to reduce the cost of disposing of radioactive waste, which is a byproduct of operating nuclear plants.[34]

Such subsidies could be viewed as compensating utilities that choose zero-emissions technologies, such as nuclear, for the potential public benefits of mitigating carbon dioxide emissions; however, such subsidies are inefficient and counterproductive in comparison to charges for emitting carbon dioxide.[35] Because production tax credits reduce the price of electricity, consumers might use electrical power less efficiently and expand the gap between the price of electricity and its cost to society at the expense of the general taxpayer. Alternatively, a charge on carbon dioxide emissions that was representative of the damage those emissions cause would equate the price of electricity to its social cost. Such a price would lead to more-efficient use of electricity, because the utility and consumer, rather than the general taxpayer, would pay for the cost of carbon dioxide emissions.

3. Results and Implications of the Analysis

Under the provisions of the Energy Policy Act of 2005, it is probable that at least a few nuclear power plants will be built over the next decade, most likely in markets where electricity usage and the corresponding demand for additional base-load capacity are expected to grow significantly. Ultimately, however, the longer-term competitiveness of nuclear technology as a source of electricity is likely to depend on policymakers' decisions regarding carbon dioxide constraints. If such constraints are implemented, nuclear power will probably enjoy a cost advantage over conventional fossil-fuel alternatives as a source of electricity-generating capacity. Today, even the anticipation that carbon dioxide emissions will be priced is a factor being weighed in investors' decisions about new base-load capacity. Those conclusions are tentative, though, because the electricity industry faces numerous uncertainties. If expectations related to future market conditions—especially those pertaining to construction costs or fuel prices—shift before investors commit to the construction of new base-load capacity, the prospects for new nuclear capacity could change dramatically.

The Outlook for Investment in the Absence of Carbon Dioxide Charges and EPAct Incentives

In the Congressional Budget Office's reference scenario, the estimated levelized costs of new capacity based on conventional coal or conventional natural gas technology are roughly equivalent. By comparison, the levelized costs of the other options under consideration are much higher. Specifically, the levelized costs for building and operating a new nuclear power plant are estimated to be about 30 percent more than the cost of either a conventional coal or natural gas plant. The costs for innovative coal and natural gas plants that capture and store carbon dioxide are even greater, exceeding those of the lowest- cost conventional fossil-fuel options by 50 percent. Accordingly, in the absence of carbon dioxide constraints and without the incentives of EPAct, utilities would probably continue to build power plants relying on conventional fossil-fuel technologies to meet increases in base-load electricity demand.

Differences in construction and fuel costs explain the differences between the levelized costs of conventional fossil- fuel technologies and nuclear technology. In the reference scenario, the cost of building a new coal plant is about two-thirds the cost of building a nuclear plant, and the cost of building a natural gas plant is even less (about a third of that required to build a nuclear plant). The levelized costs for conventional coal and natural gas technologies converge because the higher cost of building a coal plant is offset by the higher cost of using natural gas as a fuel. The highest-cost alternatives considered in the reference scenario, innovative fossil-fuel plants that capture and store carbon dioxide, are encumbered by the fact that they are more costly to build and fuel than their conventional counterparts. In the reference scenario, the levelized cost of a nuclear plant is about 15 percent below that of an innovative coal-fired facility with CCS (the least expensive of the two innovative fossil-fuel technologies).

CBO's results echo those of other studies. Researchers at the Energy Information Administration and Massachusetts Institute of Technology found that, in the absence of carbon dioxide charges and EPAct incentives, new nuclear technology would have a higher levelized cost than conventional fossil-fuel technologies. Some disagree ment exists about the size of the gap, however (see Figure 3-1).[36] The researchers at MIT predicted a larger cost gap because they concluded that an investment in new nuclear capacity would be riskier than an investment in conventional fossil-fuel technologies; consequently, they projected higher financing costs for nuclear technology.

Neither the assumptions underlying that study nor CBO's base-case assumptions explicitly included the additional costs that utilities would have to pay if not for the limited liability protection offered under the Price-Anderson Nuclear Industries Indemnity Act, a policy long in effect (and extended by EPAct) that is implicitly captured in the reference scenario's assumptions. Supplementary analysis that expands on the reference scenario by exploring the likelihood that a catastrophic nuclear accident might occur suggests that removing that insurance subsidy would probably increase the levelized cost of nuclear generation by no more than 1 percent (see Box 3-1).

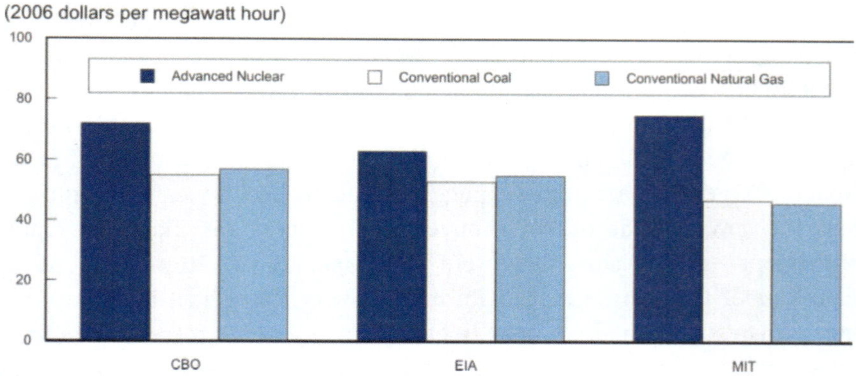

Source: Congressional Budget Office (CBO).

Notes: EIA = Energy Information Administration; MIT = Massachusetts Institute of Technology.

Electricity-generating capacity is measured in megawatts; the electrical power generated by that capacity is measured in megawatt hours. During a full hour of operation, 1 megawatt of capacity produces 1 megawatt hour of electricity, which can power roughly 800 average households.

Advanced nuclear technology refers to third-generation reactors. Conventional coal power plants are assumed to use pulverized coal technology, which produces energy by burning a crushed form of solid coal. Conventional natural gas power plants are assumed to convert gas into electricity using combined-cycle turbines.

CBO's estimates of levelized costs presented here are based on a reference scenario that excludes both the effects of prospective carbon dioxide constraints and the impact of incentives provided by the Energy Policy Act of 2005. EIA's levelized costs are based on its reference case, which includes the assumption that power plants will be built in 2015. MIT's levelized costs derive from its base case, which assumes a 40-year capital recovery period and an 85 percent capacity factor. MIT's results were converted to 2006 dollars using the gross domestic product price index and do not include the cost of delivering electricity.

Figure 3-1. Levelized Cost of Electricity in Comparable Studies

The Outlook for Investment under Carbon Dioxide Charges

Putting in place a cap-and-trade system that limited the amount of carbon dioxide power plants could emit or levying a tax on such emissions would encourage the construction of new nuclear capacity by increasing the cost of generating electricity with competing fossil-fuel technologies. Carbon dioxide constraints would have no direct effect on the levelized cost of nuclear plants and would have only a small effect on innovative fossil-fuel plants with CCS technology (which are assumed to capture 90 percent of carbon dioxide emissions). In general, under the assumptions incorporated in CBO's analysis, nuclear technology would become a more competitive source of new base-load capacity as the cost of emitting carbon dioxide increased. Eventually, if carbon dioxide charges became high enough, it would be economical for utilities to construct new nuclear or new conventional natural gas plants to replace conventional coal plants that were still operational. Regardless of the magnitude of the charges that might be imposed, however, it is unlikely that nuclear plants could be built quickly enough at any point in the near future to replace existing coal power plants, which currently account for half of the electricity generated in the United States.

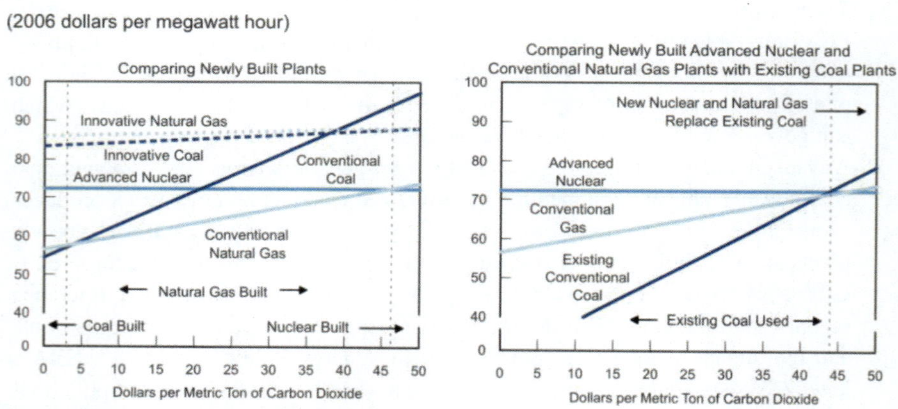

Source: Congressional Budget Office.

Notes: Electricity-generating capacity is measured in megawatts; the electrical power generated by that capacity is measured in megawatt hours. During a full hour of operation, 1 megawatt of capacity produces 1 megawatt hour of electricity, which can power roughly 800 average households.

Advanced nuclear technology refers to third-generation reactors. Conventional coal power plants are assumed to use pulverized coal technology, which produces energy by burning a crushed form of solid coal. Conventional natural gas power plants are assumed to convert gas into electricity using combined-cycle turbines. Innovative coal and natural gas technologies are assumed to capture and store most carbon dioxide emissions.

Figure 3-2. Levelized Cost of Alternative Technologies to Generate Electricity under Carbon Dioxide Charges

Would Nuclear Technology Be the Choice for Expanding Capacity?

Adding a carbon dioxide charge of about $45 per metric ton to the levelized cost estimates in the reference scenario would make nuclear power the least expensive source of additional base-load capacity (see the left panel of Figure 3-2). Up to that threshold, at all but the lowest level of charges, conventional natural gas technology would probably be the least costly option. Because coal is more carbon-intense than natural gas, the cost advantage of new capacity based on natural gas technology would grow in relation to coal technology as carbon dioxide charges increased; but the advantage that natural gas technology enjoyed over nuclear technology would shrink and eventually disappear as emission charges reached about $45 per metric ton. Thereafter, the levelized cost advantage of nuclear technology over conventional gas technology would grow. Although carbon dioxide charges would not change the cost of nuclear power plants at all, they would increase the cost of innovative fossil-fuel alternatives; as a result, the cost advantage that nuclear technology held over those technologies would increase with carbon dioxide charges but at a slower rate than that observed with conventional fossil-fuel technologies.

BOX 3-1. THE COST OF LIABILITY FOR NUCLEAR ACCIDENTS

Among its various provisions, the Energy Policy Act of 2005 extended the Price-Anderson Nuclear Industries Indemnity Act, which limits the industry's liability for accidents at nuclear power plants. In practice, Price-Anderson subsidizes utilities by reducing their cost of carrying liability insurance. Instead of purchasing full coverage, operators of nuclear power plants are required to obtain coverage only up to the liability limit, which is currently set at about $10 billion per accident.[1] The value of the subsidy is the difference between the premium for full coverage and the premium for $10 billion in coverage. On the basis of data obtained from two studies—one conducted by the Nuclear Regulatory Commission (NRC) and the other by the Department of Energy (DOE)—the Congressional Budget Office (CBO) estimates that the subsidy probably amounts to less than 1 percent of the levelized cost for new nuclear capacity.[2]

To assess the health hazards that existing nuclear power plants could pose, analysts at the NRC estimated the probability of radioactive releases occurring at several nuclear facilities, including the Surry power station in

Virginia, and the consequence of such an event.[3] Damage to property and possible injury or loss of life caused by a hypothetical accident at that facility could be pertinent to assessing the liability of proposed nuclear plants because several of them would be located in areas of the Southeast with roughly similar population densities. For the Surry power station, the NRC study provides assessments of both internally initiated accidents (which could be caused by malfunctioning equipment or human error) and externally initiated accidents (which could result from a fire or earthquake). According to the study, an internally initiated accident at such a facility that on average caused more than 10 deaths would occur, at most, once every million years. A fire-related accident causing more than 1,000 deaths on average would occur, at most, once every million years. CBO's analysis adopted those probabilities and results for the sake of determining liability from fatalities. To that, CBO added estimates of injuries and property damage to provide a more complete estimate of liability.

CBO based its assessment of liability from injuries and property damage on the DOE report, which modeled a radioactive release at the Limerick facility near Philadelphia. That scenario includes, in addition to the number of fatalities, estimates of injury and property damage, from which CBO inferred potential liability resulting from an accident at the Surry plant.

On the basis of the probability of fatal accidents estimated in the NRC report and the estimates of damage from such accidents in the DOE report, it appears that catastrophic accidents are possible but likely to be rare; CBO estimates that an accident causing about $500 billion in damages will occur an average of 3 out of every 100 million years.[4] Because such potential damages are spread over a long period, the long-run average of damages per year (the expected cost) would be only about $600,000. That figure does not include the cost of nonfatal accidents, which might already be covered by the $10 billion in damages for which the nuclear power industry is held liable under the Price-Anderson Act. If so, the projected annual subsidy is about $600,000 per reactor as well.

Insurance premiums represent a small portion of the levelized cost for a nuclear power plant. Even if the analysis based on the Surry facility understates the expected cost of fatal nuclear accidents by a factor of 10, paying a fair premium would not lead to large changes in the levelized cost. In CBO's reference scenario, increasing the insurance premium by $6 million per year increases the levelized costs by 1 percent.

[1] That $10 billion in coverage has two layers: The owner of a nuclear plant is required to purchase primary insurance covering liability up to $300 million. In the event of an accident, liability for damages assessed at between $300 million and $10 billion would then be shared among the owners of all U.S. nuclear plants, who would pay a "retroactive premium."

[2] See Nuclear Regulatory Commission, *Severe Accident Risks: An Assessment for Five U.S. Nuclear Power Plants*, NUREG1150 (December 1990); and Department of Energy, *Technical Guidance for Siting Criteria Development*, SAND-81-1549 (December 1982). CBO's estimate was derived to evaluate the sensitivity of levelized costs (or the minimum price of electricity at which a technology generates enough revenue to be economically viable) to limits on liability but should not be interpreted as a precise estimate of the expected cost of liability.

[3] A description and evaluation of the NRC's probabilistic risk assessment models is provided in *Nuclear Power Joint Fact- Finding* (Keystone Center, June 2007).

[4] Each fatality is assumed to lead to $5,000,000 in liability, and each injury is assumed to cause $2,500,000 in liability.

Variations in the base-case assumptions about the costs for construction, fuel, or financing could increase the levelized cost of meeting base-load demand with nuclear technology relative to that for alternative technologies. For example, if the construction costs of all generation technologies doubled, carbon dioxide charges would have to be set at $150 per metric ton for nuclear technology to be preferred over conventional natural gas technology. (The concluding section of this chapter provides a more comprehensive assessment of the sensitivity of various technologies to variations in the base-case assumptions.)

Would New Nuclear Plants Replace Any Existing Coal Capacity?

When carbon dioxide charges are added to the levelized costs estimated in the reference scenario, they fall most heavily on coal technologies. The effect is so significant that, at a carbon dioxide charge of about $45 per metric ton of emissions, this analysis suggests that utilities could build and operate new nuclear or conventional natural gas plants at a lower levelized cost than continuing to operate existing coal plants. But whether such a switch would occur would depend on other factors as well— including the markets for the components and labor necessary to build new reactors and the market for natural gas.

Even if carbon dioxide charges over $45 per metric ton were implemented, it would take decades for sufficient nuclear capacity to be put in place before most utilities could consider substituting new nuclear capacity for existing coal plants. Replacing the 300,000 megawatts of existing coal capacity would require hundreds of new nuclear plants. The capacity of the industry that builds nuclear plants and its suppliers of components is currently constrained and unlikely to expand rapidly enough for even tens of plants to be built in the next decade. For example, the Brattle Group (a consulting firm) has pointed out that the skilled labor necessary to erect power plants is in short supply and could be slow to expand if a surge in the demand for nuclear plants occurred.[37] Also, the supply of steel forgings necessary to build a reactor's containment vessel—a structure that prevents radiation from leaking into the atmosphere—is limited.[38]

Although the trend toward natural gas technology that was evident in the 1 990s could always recur, it is not likely that natural gas technology would completely replace coal technology as a source of electrical power. The primary reason is that increased demand for natural gas would exert upward pressure on the price of that fuel, perhaps pushing costs above the levels included in the reference scenario. To illustrate, at the highest prices for natural gas considered in CBO's analysis of market and policy uncertainties, utilities would be extremely unlikely to prefer natural gas to either existing coal plants or new nuclear plants.

The Outlook for Investment under the Energy Policy Act of 2005

The maximum allocation of benefits currently available under EPAct would most likely lead to the planning and construction of at least a few new nuclear plants in the next decade, even in the absence of carbon dioxide charges. (Table 1-1 on page 10 provides a complete list of incentives created or extended by EPAct.) If just a few nuclear plants qualified for the incentives, the most substantial one—the production tax credit—would lead to sizable reductions in those plants' corporate income tax liability during the first several years of operation. Nuclear projects eligible for federal loan guarantees, which cover up to 80 percent of construction costs, would benefit from reductions in financing costs. The preferential tax treatment of decommissioning funds— funds that utilities are required to set aside to cover the cost of safely shutting down and securing a nuclear plant at the end of its useful life—would provide far less financial incentive because the discounted present value of the cost of decommissioning is small. (Although the decommissioning of a 1,350-megawatt plant costs nearly $500 million, by CBO's estimate, the present value of that cost would be much smaller because that sum would be spent 40 years after the power plant was constructed. In the absence of the preferential income tax rate, decommissioning costs would still account for less than 1 percent of the levelized cost of generating electricity with new nuclear capacity.)

The value of other EPAct incentives—including cost-sharing for the licensing and design of advanced reactors, which is offered by the Nuclear Power 2010 program, and the protection afforded by delay insurance, which insures investors in new nuclear plants for the financial risk caused by litigation or licensing delays, is not reflected in CBO's analysis of EPAct incentives because those subsidies directly reduce the cost of only the first plants built of any new type. Those first-of-a-kind costs are not projected to have large effects on the levelized cost estimates. (The value of both the Nuclear Power 2010 program and delay insurance is discussed further in Box 3-2.)

Projects Receiving the Maximum Benefits under EPAct

When the levelized costs estimated under the reference scenario are changed to account for the benefits provided by EPAct, nuclear technology emerges as a competitive source for a limited amount of new capacity, with costs roughly comparable to those of additional capacity based on conventional fossil-fuel technologies. Accounting for the EPAct incentives also reduces the levelized cost of both innovative fossil-fuel options. The levelized cost of innovative natural gas plants falls by about 5 percent, and that of innovative coal plants decreases by 20 percent. However, those technologies are still more costly than conventional fossil-fuel alternatives or nuclear technology. (See Figure 3-3 for an illustration of the cost of each technology under the full provision of EPAct incentives.) After those first few new nuclear plants qualified for EPAct incentives, the cost of new nuclear capacity would exceed the cost of new conventional coal capacity.

Production Tax Credits

Production tax credits provided under EPAct reduce by almost 15 percent the levelized cost of nuclear technology estimated in the reference scenario, making them the incentive with the greatest potential value to investors.

The effect of the production tax credits on levelized costs would be smaller if more than 6,000 megawatts of qualified nuclear capacity (equivalent to the output of three to five plants) was constructed; but construction of more capacity would indicate that nuclear technology did not require the full allotment of credits to be commercially viable.

Box 3-2. The Value of the Nuclear Power 2010 Program and Delay Insurance

The Energy Policy Act of 2005 (EPAct) includes provisions authorizing the Nuclear Power 2010 program and Standby Support, a program offering insurance against regulatory delays. Both are intended to encourage investment in advanced nuclear technology by covering a share of "first-of-a-kind" (FOAK) costs. Specifically, the Nuclear Power 2010 program shares the cost of licensing and designing new nuclear power plants, and the delay insurance helps mitigate risks that are particular to the first plants to test the Nuclear Regulatory Commission's revised licensing process.

Through fiscal year 2007, the Nuclear Power 2010 program contributed roughly $280 million to funds spent by three industry consortia that were attempting to design and obtain certification for the first power plants using advanced nuclear technology.[1] Such design and licensing costs for each new plant are estimated to be $300 million to $500 million. However, because utilities that decided to build subsequent plants using the same design would benefit from the knowledge and experience gained during the construction of the first few plants, their costs would be less, roughly $100 million.[2] The costs for subsequent plants are the basis of the design and licensing costs used in CBO's reference scenario, where they account for 5 percent of a nuclear power plant's levelized cost.

If the additional $200 million to $400 million in first-of-a-kind design and licensing costs was added to a single plant in CBO's reference scenario, the levelized cost of that initial nuclear plant would increase by roughly 15 percent. But the first utilities to build plants might be able to share some FOAK costs by forming consortia, even in the absence of the federal program. Accordingly, in the absence of federal support for design and licensing costs, levelized costs for the first plants might increase by only a fraction of the 15 percent.

Even if utilities pay the Department of Energy's (DOE's) estimated subsidy costs for the delay insurance, the program would reduce utilities' cost of generating electricity by the transferring financial risk from private investors to taxpayers. However, according to an assessment by DOE, the amount of financial risk transferred is small in comparison to that of the department's program providing guarantees for construction loans.[3] For the delay insurance, DOE estimates a hypothetical subsidy cost of $14 million for a reactor with a capacity of 1,090 megawatts. Under the terms of the insurance, the Treasury would reimburse a utility for up to $500 million in costs in the event of a covered delay.[4] For the loan guarantee program, the maximum reimbursement is roughly 10 times as much.

[1] Each consortium is developing a standard nuclear power plant blueprint based on a different third-generation reactor design. One of the three reactor designs, the advanced boiling water reactor, received final approval before the Nuclear Power 2010 program was initiated, but the program supports the design and licensing of a power plant's components surrounding such a reactor.

[2] See Louis Long and others, *A Roadmap to Deploy New Nuclear Power Plants in the United States by 2010* (prepared for the Department of Energy, October 2001).

[3] See Department of Energy, "Standby Support for Certain Nuclear Plant Delays; Final Rule," *Federal Register*, vol. 71, no. 155 (August 2006), p. 46324.

[4] The $500 million in coverage is available only to the first two reactors. The next four reactors are eligible for reduced coverage. DOE estimates that the subsidy cost for the reduced coverage is roughly half of the $14 million subsidy cost for the first two reactors.

Because the credits would not be used until after a plant began operating—in other words, once electricity had been sold and the utility had incurred sufficient tax liability—the reduction in the levelized cost of generating electricity for qualifying nuclear plants would necessarily be less than the nominal value of the credits awarded to each project. Thus, even though credits of $18 per megawatt hour of generated electricity are equal to about one- quarter of the levelized cost estimated in the reference scenario, after discounting, the credits would reduce the cost of nuclear capacity by only about 15 percent if they were used within three years of being awarded.

Loan Guarantees

The maximum coverage available under the loan guarantee program—a guarantee on debt covering 80 percent of a plant's construction costs, which implies that investors' equity would cover the remaining 20 percent—would most likely reduce the levelized cost of new nuclear capacity by about 10 percent. But not all prospective nuclear plants would necessarily receive a guarantee of debt covering 80 percent of construction costs because the criteria for qualifying are restrictive. The Department of Energy has indicated that it will deny a utility's application for a loan guarantee if the project is not deemed to be both innovative (essentially, in the case of nuclear technology, a plant design that has not been built in the United States) and commercially viable, and that no more than three plants based on each advanced reactor design can be considered innovative. The 30 plants currently being proposed use five reactor designs, so at most, 15 of those plants would qualify as innovative. In addition, just because a plant is considered both innovative and commercially viable does not mean it will receive the maximum guarantee of 80 percent. Under the base-case assumptions, covering 80 percent of construction costs would require guaranteeing debt with a face value of $4.5 billion to $7.5 billion for each plant (depending on the size of the reactor). Providing the maximum coverage to three plants based on each of the five reactor designs would result in roughly $100 billion in loan guarantees, a commitment that has not been proposed, let alone funded. (The President's budget proposed a limit of $18.5 billion [in nominal dollars] on the cumulative amount of loan guarantees for new nuclear plants over the 2008–2011 period.)[39] The loan guarantee program could encourage investors to choose relatively risky projects over more certain alternatives because they would be responsible for only about 20 percent of a project's costs but would receive 100 percent of the returns that exceeded costs.[40]

Incentives and Impediments at the State and Local Levels

Because some states and localities regulate the rates that consumers pay for electricity or offer incentives for specific technologies, the levelized cost of nuclear technology in certain areas of the country could be lower than the estimates in this analysis. Other states have policies that deter investment in new nuclear or coal capacity altogether, which renders the levelized cost of the prohibited technology moot.

States and localities encourage investment in new nuclear capacity through a variety of policies. Over half of the currently proposed new nuclear plants are sited in southeastern states, where most electricity-generation capacity is owned by utilities that charge regulated rates. To the extent that rate regulation guarantees that customers will reimburse utilities for the cost of building a new plant, financial risk is transferred from investors to customers, which leads to larger reductions in the cost of capital- intense technologies such as nuclear. In several of those states, additional incentives that could further reduce the cost of nuclear power are under consideration. Those provisions include allowing higher rates of return for nuclear power than for other technologies, allowing utilities to recover some construction costs before plants begin operations, and tax incentives. State incentives for new nuclear power plants are not limited to states with traditional regulation in place. For instance, Texas, a state that allows markets a large role in setting electricity prices, has expanded a tax incentive initially designed to encourage investment in renewable energy technologies to apply to new nuclear capacity. Last, California and a number of eastern states are considering legislation that would limit carbon dioxide emissions, which could increase the competitiveness of nuclear and innovative fossil-fuel technologies. As of 2007, however, the only states in that group that had proposed sites for new nuclear power plants were Maryland, Pennsylvania, and New York.

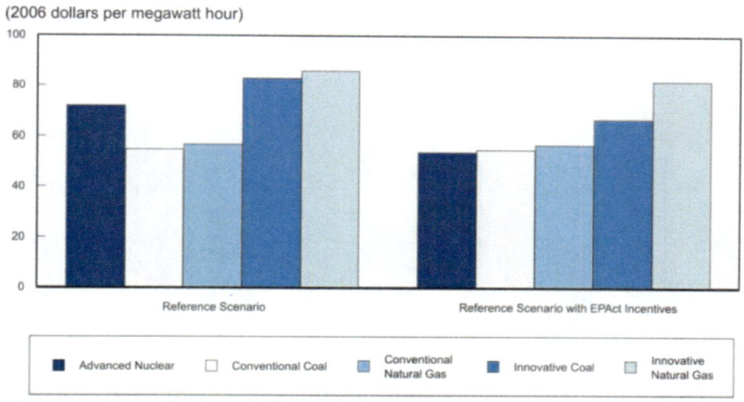

Source: Congressional Budget Office (CBO).

Notes: CBO's reference scenario excludes both the effects of prospective carbon dioxide constraints and the impact of incentives provided by the Energy Policy Act of 2005 (EPAct). The estimate of the effect of EPAct incentives assumes the maximum production tax credits, loan guarantees, and investment tax credits. The production tax credits are shared among 6,000 megawatts or less of advanced nuclear capacity. Advanced nuclear and innovative fossil fuel technologies receive loan guarantees covering 80 percent of construction costs. Innovative coal technology receives investment tax credits for 20 percent of construction costs.

Electricity-generating capacity is measured in megawatts; the electrical power generated by that capacity is measured in megawatt hours. During a full hour of operation, 1 megawatt of capacity produces 1 megawatt hour of electricity, which can power roughly 800 average households.

Advanced nuclear technology refers to third-generation reactors. Conventional coal power plants are assumed to use pulverized coal technology, which produces energy by burning a crushed form of solid coal. Conventional natural gas power plants are assumed to convert gas into electricity using combined-cycle turbines. Both innovative coal and innovative natural gas technologies are assumed to capture and store most carbon dioxide emissions.'

Figure 3-3. Levelized Cost of Alternative Technologies to Generate Electricity With and Without EPAct Incentives

At least 11 states have prohibitions against the construction of new nuclear facilities until certain provisions governing the long-term disposal of spent nuclear fuel are put in place.[41] Minnesota completely bans the construction of new nuclear power plants.

Other prohibitions apply to conventional coal technology. A California law essentially prohibits the construction of any new coal-fired power plant that does not employ CCS technology. In New England, utilities have been blocked from building new coal-fired plants for over a decade.

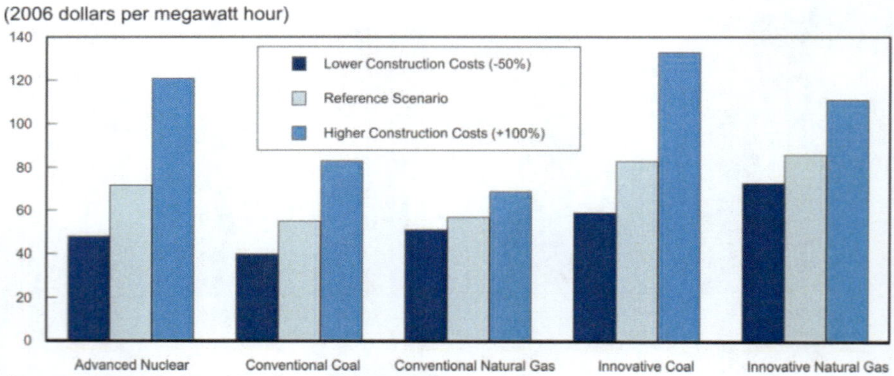

Source: Congressional Budget Office (CBO).

Notes: Electricity-generating capacity is measured in megawatts; the electrical power generated by that capacity is measured in megawatt hours. During a full hour of operation, 1 megawatt of capacity produces 1 megawatt hour of electricity, which can power roughly 800 average households.

CBO's reference scenario excludes both the effects of prospective carbon dioxide constraints and the impact of incentives provided by the Energy Policy Act of 2005.

CBO calculated levelized costs for the reference scenario using estimates of overnight construction costs from the Energy Information Administration (EIA). **In the case of "Lower Construction Costs," CBO halved EIA's estimates and recalculated levelized costs on that basis. In the case of "Higher Construction Costs," CBO doubled EIA's estimates and recalculated the costs.**

Advanced nuclear technology refers to third-generation reactors. Conventional coal power plants are assumed to use pulverized coal technology, which produces energy by burning a crushed form of solid coal. Conventional natural gas power plants are assumed to convert gas into electricity using combined-cycle turbines. Both innovative coal and innovative natural gas technologies are assumed to capture and store most carbon dioxide emissions.

Figure 3-4. Sensitivity of Levelized Costs to Future Construction Costs

Future Market and Policy Uncertainties

The commercial viability of new nuclear capacity depends on investors' perceptions of future market conditions and carbon dioxide constraints when investment decisions are finalized. Licensing and regulatory approval for building new nuclear plants in the United States are expected to take about three years, so the construction of the first new nuclear plants would be unlikely to start until 2010 at the earliest. At that point, the commercial viability of a new plant would depend on anticipated market conditions and policy outcomes over the operating life of the plant, which may exceed 40 years. A combination of factors—recent volatility in construction costs and natural gas prices, nuclear power's history of construction cost overruns, and uncertainty about future policy on carbon dioxide emissions—indicates that a wide range of costs are plausible for each of the technologies considered. Those ranges demonstrate that the future competitiveness of each technology and thus the conclusions presented in this analysis are quite uncertain.

Costs under Alternative Market Conditions

If the base-case assumptions incorporated in CBO's reference scenario did not hold—for instance, if construction costs for new nuclear power plants proved to be as high as the average cost of nuclear plants built in the 1 970s and 1980s or if natural gas prices fell back to the average levels seen in the 1990s—new nuclear capacity would be an unattractive investment, regardless of the incentives provided by EPAct. Specifically, CBO compared the levelized cost of electricity from new capacity assuming that the overnight construction costs for nuclear technology would be 25 percent higher or that the fuel costs for natural gas capacity would be 50 percent lower than in the base case. Nuclear technology benefiting from EPAct incentives was about 15 percent more expensive than conventional fossil-fuel capacity in the first case and about 50 percent more expensive than conventional natural gas capacity in the second case. However, such variations in construction or fuel costs would be less likely to deter investment in new nuclear capacity if investors anticipated future charges for emitting carbon dioxide. New nuclear capacity could even compete at lower carbon dioxide charges if the price of natural gas continued to rise or if the construction cost reductions predicted by reactor designers were accurate (and thus below the costs assumed in the reference scenario).

Construction Costs. To examine the effect of uncertainties related to overnight construction costs, CBO calculated the levelized cost of each of the technological alternatives using values for construction costs that were 50 percent less than and 100 percent more than the assumptions in the reference scenario (see Figure 3-4). In the reference scenario, CBO assumed an overnight construction cost for new nuclear capacity of about $2.4 million per megawatt, so in the "lower" and "higher" cases, CBO used a cost of about $1.2 million and $4.8 million, respectively. Taking into consideration the history of very large construction cost overruns that have plagued the nuclear power industry in the past, the high end of the range encompasses costs well above those included in the reference scenario. Utilities would be unlikely to invest in new nuclear plants if construction costs for nuclear plants were twice those assumed in the reference scenario, as the levelized cost of a nuclear plant would be well over $100 per megawatt hour.

However, the construction costs for new capacity using the other technologies are subject to uncertainty as well. Adjusted for general inflation, construction costs for new power plants increased by 15 percent between 1994 to 2006, with most of that increase occurring over the past three years.[42] If that trend continued, the overnight costs assumed in the reference scenario for all of the technologies would be too low. If the construction costs of each technology increased by a similar percentage, the impact on the levelized cost of the technologies with the highest overnight costs in the reference scenario—nuclear and innovative coal—would be the greatest. Conventional natural gas technology would become less expensive by comparison because construction costs account for a smaller percentage of that technology's levelized cost. Conversely, reductions in construction costs would have a disproportionately large effect in reducing the relative cost of those technologies that had high construction costs in the reference scenario.

Fuel Costs. The price of the primary fuels used by each of the base-load technology options has increased in recent years. Contracted uranium prices (the prices that operators of nuclear plants pay for most of their fuel) increased by 40 percent between 2004 and 2006. Spot prices for uranium (the prices at which a relatively small amount of uranium trades on commodity exchanges) have climbed steeply and then fallen over the past year. Natural gas prices have increased dramatically (see Figure 2-1 on page 20), and even spot prices for coal have recently increased to levels that are well above long-run averages. To capture the effects of uncertainty surrounding fuel prices, CBO estimated the levelized cost of each of the five technology options on the basis of fuel prices set at levels 50 percent less than and 100 percent more than those assumed in the reference scenario. (The results are displayed in Figure 3-5.)

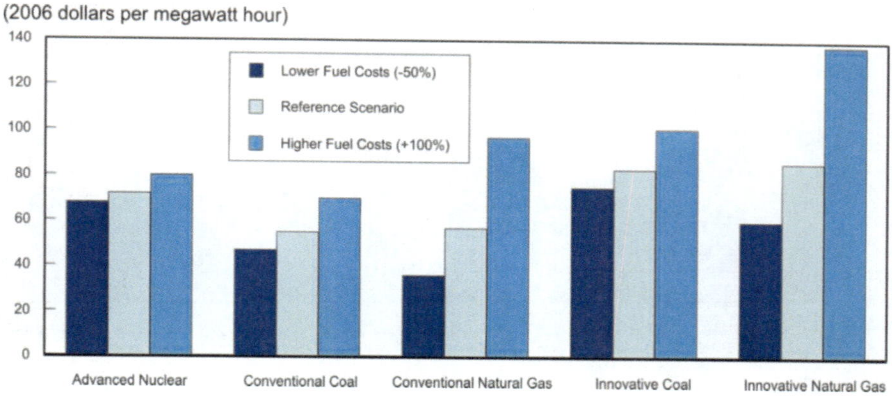

Source: Congressional Budget Office (CBO).

Notes: Electricity-generating capacity is measured in megawatts; the electrical power generated by that capacity is measured in megawatt hours. During a full hour of operation, 1 megawatt of capacity produces 1 megawatt hour of electricity, which can power roughly 800 average households.

CBO's reference scenario excludes both the effects of prospective carbon dioxide constraints and the impact of incentives provided by the Energy Policy Act of 2005.

CBO estimated levelized costs for the reference scenario using estimates of fuel costs from the Energy Information Administration (EIA). **In the case of "Lower Fuel Costs," CBO halved EIA's estimates and recalculated levelized costs on that basis. In the case of "Higher Fuel Costs," CBO doubled EIA's estimates and** recalculated the costs.

Advanced nuclear technology refers to third-generation reactors. Conventional coal power plants are assumed to use pulverized coal technology, which produces energy by burning a crushed form of solid coal. Conventional natural gas power plants are assumed to convert gas into electricity using combined-cycle turbines. Both innovative coal and innovative natural gas technologies are assumed to capture and store most carbon dioxide emissions.

Figure 3-5. Sensitivity of Levelized Costs to Future Fuel Costs

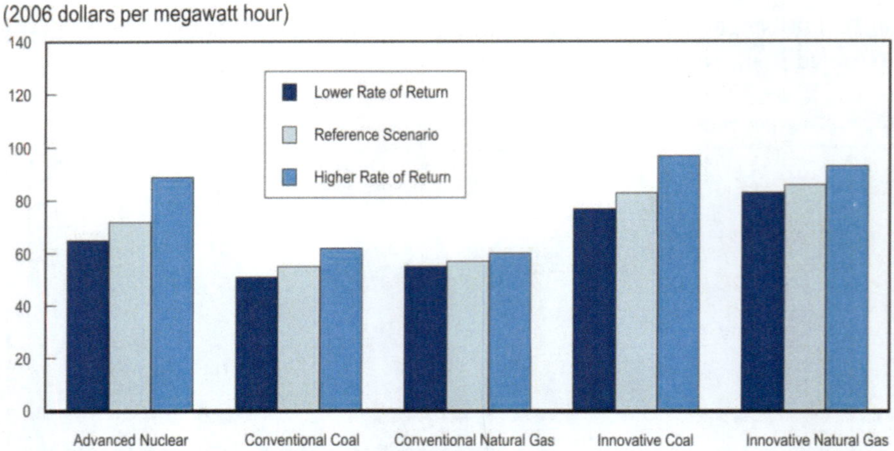

Source: Congressional Budget Office (CBO).

Notes: Electricity-generating capacity is measured in megawatts; the electrical power generated by that capacity is measured in megawatt hours. During a full hour of operation, 1 megawatt of capacity produces 1 megawatt hour of electricity, which can power roughly 800 average households.

CBO's reference scenario excludes both the effects of prospective carbon dioxide constraints and the impact of incentives provided by the Energy Policy Act of 2005.

CBO estimated levelized costs for the reference scenario using a financing rate of about 10 percent. For the category "Lower Rate of Return," CBO recalculated levelized costs using a financing rate of about 8-^{3}/4 percent. For the category "Higher Rate of Return," CBO recalculated those costs using a rate of about 12-^{1}/2 percent. In each case, financing rates were derived from CBO's assumptions for debt and equity financing (which are discussed in detail in the Web supplement to this paper) and rounded to the nearest quarter of a percentage point.

Advanced nuclear technology refers to third-generation reactors. Conventional coal power plants are assumed to use pulverized coal technology, which produces energy by burning a crushed form of solid coal. Conventional natural gas power plants are assumed to convert gas into electricity using combined-cycle turbines. Both innovative coal and innovative natural gas technologies are assumed to capture and store most carbon dioxide emissions.

Figure 3-6. Sensitivity of Levelized Costs to Future Financing Risks

The analysis shows that assumptions about fuel prices have particularly large effects on the estimated levelized cost of natural gas technologies. For conventional natural gas, doubling the cost of fuel used in the reference scenario increases the levelized cost of that technology by over 70 percent. For innovative natural gas plants, the corresponding increase in levelized cost is just over 60 percent. By contrast, nuclear power has the lowest fuel cost of the technologies considered under the base-case assumptions. Doubling that cost would increase the levelized cost of new nuclear capacity by about 15 percent above that assumed in the reference scenario.

Financing Costs. A levelized cost analysis accounts for risk by assuming that investors will require a higher rate of return for riskier projects. The effect of uncertainty on the return that investors will require to finance new base-load capacity can be shown for all of the options by calculating the levelized cost of each option at a lower rate of return (8-$^{3}/4$ percent) and at a higher return (12-$^{1}/_{2}$ percent) and comparing those results with the rate of return (10 percent) used in the reference scenario. (The results of that analysis are presented in Figure 3-6.)

The changes in levelized costs follow the same pattern as that produced under higher construction costs. The levelized costs for nuclear and innovative fossil-fuel technologies increase by more than those for the conventional fossil-fuel options, which require smaller up-front investments. But some observers rate the level of risk attached to nuclear technology as somewhat higher than that of the other alternatives, **independent of the technology's** relatively high construction costs. The prospect that investors would require a higher rate of return relative to that of other options to compensate for a risk unique to nuclear power (for instance, an accident at an existing nuclear plant could significantly delay the construction of new plants, as was the case after the Three Mile Island accident) can be assessed by comparing the levelized cost of nuclear technology at the highest rate of return with the levelized cost of the other options calculated for the reference scenario. For example, the levelized cost of nuclear technology calculated at a 12-$^{1}/_{2}$ percent real rate of return is 65 percent higher than the levelized cost of conventional coal with a 10 percent real rate of return. With the same comparison applied to nuclear technology the least expensive innovative fossil-fuel technology—coal with CCS—the levelized cost of nuclear technology exceeds that of innovative coal technology by about 5 percent. Such comparisons suggest that if financial markets required a substantially higher rate of return for new nuclear technology than for other base-load technologies, investment in new nuclear plants would not be commercially viable.

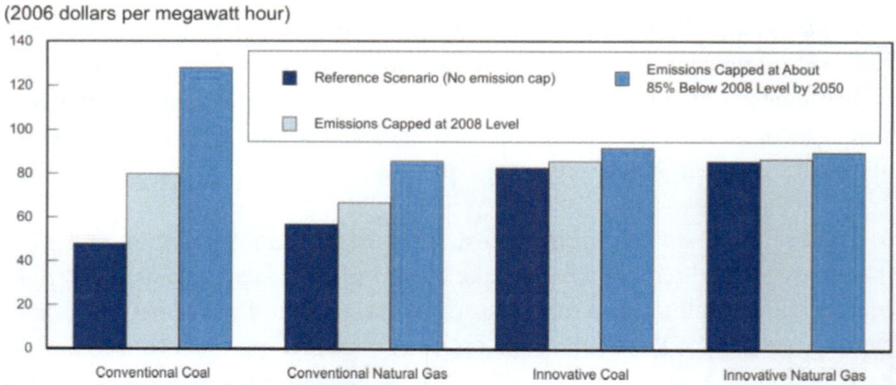

Source: Congressional Budget Office.

Notes: Electricity-generating capacity is measured in megawatts; the electrical power generated by that capacity is measured in megawatt hours. During a full hour of operation, 1 megawatt of capacity produces 1 megawatt hour of electricity, which can power roughly 800 average households.

CBO's reference scenario excludes both the effects of prospective carbon dioxide constraints and the impact of incentives provided by the Energy Policy Act of 2005.

In the reference scenario, carbon dioxide emissions are not constrained, so they are not priced. In the second case, the number of allowances issued each year for emitting carbon dioxide would correspond to a cap at roughly the level of emissions in 2008. In the third case, the number of such allowances would correspond to a cap about 85 percent below that level by 2050. The allowance prices resulting from those hypothetical constraints are listed in the Web supplement to this paper.

Conventional coal power plants are assumed to use pulverized coal technology, which produces energy by burning a crushed form of solid coal. Conventional natural gas power plants are assumed to convert gas into electricity using combined-cycle turbines. Both innovative coal and innovative natural gas technologies are assumed to capture and store most carbon dioxide emissions.

Figure 3-7. Sensitivity of Levelized Costs to Potential Carbon Dioxide Constraints

Costs under Prospective Carbon Dioxide Constraints

It is less likely that the cost of building and fueling conventional coal plants will vary toward the extremes of the ranges included in this analysis than it would for the other generation options considered. Construction costs for coal plants have been less volatile than for nuclear plants, and the abundance of U.S. coal supplies has historically led to relatively stable prices for coal. Still, investors in new coal capacity face substantial uncertainty because of the prospect that carbon dioxide constraints will be implemented and the variability in the prospective stringency of such constraints. Although the implications of such stringency can be observed by adding specific carbon dioxide charges to the levelized costs estimated for the reference scenario, the effect is more clearly demonstrated by comparing the levelized cost of the technology alternatives at specific levels of carbon emissions that might be targeted by policymakers. CBO compared the levelized cost of the different base-load technologies under the reference scenario with the levelized cost at two different levels of stringency, one holding future emissions at their 2008 level and the other reducing future emissions even more, to about 85 percent below their 2008 level by 2050.[43] Under the less stringent cap, the costs for electricity from newly built conventional coal and natural gas capacity would increase by about 70 percent and 20 percent, respectively. Under the more stringent cap, those costs would increase by about 165 percent and 65 percent, respectively. (Figure 3-7 shows the levelized cost under the hypothetical cap-and-trade programs.)

End Notes

[1] A cap-and-trade program would require utilities or other entities to hold permits, or allowances, to emit carbon dioxide. Because the permits would be tradable, a utility could buy them if it exceeded the emission cap or sell them if it emitted less than the cap allowed. The price at which those allowances traded would be the price of emitting carbon dioxide.

[2] The Energy Policy Act (Public Law 109-58) was signed into law on August 8, 2005. The Energy Independence and Security Act (P.L. 110-140), which was enacted in December 2007, did not provide additional incentives for nuclear technology or alter the EPAct incentives analyzed in this study.

[3] Electricity-generating capacity is commonly distinguished as base-load (that which is operated continuously, unless a plant is undergoing maintenance or repairs) or peak (that which is operated only during periods of high demand).

[4] This study considers only stack emissions of carbon dioxide— emissions resulting directly from the operation of the power plant. All technologies cause additional emissions from the construction and decommissioning of a power plant, as well as from the production of fuel.

[5] The electrical power industry's ability to continue to meet demand is conventionally measured by comparing peak usage—the amount of capacity used when electricity demand is greatest—to the amount of capacity available during the periods of peak usage. Both peak usage and the amount of capacity available to meet it increased by roughly 30 percent between 1994 and 2006, indicating the amount of excess capacity has, for the most part, stayed stable. (Peak usage data are available from EIA at *www.eia.doe.gov/cneaf/electricity/ epa/epat3p2.html* under the label "net internal demand.") However, the infrastructure for transmitting electricity between certain areas of the country is limited; accordingly, excess capacity in one region may not be available to all other regions.

[6]. Combined-cycle gas turbines generate electricity in two sequential processes, first using the energy produced by burning natural gas and then harnessing residual steam heat. The single-cycle process is generally considered an inefficient method of generating base-load power because it produces less electricity from a given amount of fuel.

[7] In this study, a nuclear power plant is defined as having one reactor; for example, if a utility built two reactors at the same site, that configuration would be considered two additional power plants.

[8] See Energy Information Administration, *U.S. Household Electricity Report* (July 2005). Electricity-generating capacity is measured in megawatts; the electrical power generated by that capacity is measured in megawatt hours. During a full hour of operation, 1 megawatt of capacity generates 1 megawatt hour of electricity, which can power roughly 800 average households.

[9] Prospects for CCS technologies and other forms of carbon sequestration are discussed in Congressional Budget Office, *The Potential for Carbon Sequestration in the United States* (September 2007).

[10] The levelized cost estimates calculated in this study do not include the cost of distributing electricity so the relevant price against which costs are compared is the wholesale price of electricity. Nor do the calculations include any impediments, incentives, or rate regulation provided by state or local governments. Several states have official or de facto prohibitions against the construction of additional nuclear power plants or additional conventional coal-fired power plants. Other states and localities encourage additional nuclear capacity through tax incentives or regulations that allow higher returns.

[11] See Energy Information Administration, *Annual Electric Power Industry Report* (2005), available at www.eia.doe.gov/cneaf/ electricity/wholesale/wholesalet2.xls.

[12] That range is based on the 10-year average of historical spreads for debt rated from B to BBB.

[13] Nuclear plant operators are required to set aside funds to cover the cost of decommissioning—that is, safely shutting down a nuclear reactor at the end of its useful life

[14] Under Internal Revenue Service guidelines for the production tax credit, once the amount of qualified nuclear capacity exceeded 6,000 megawatts, a fixed amount of total credits would be divided among all eligible capacity.

[15] The results of alternative financing assumptions are given in Chapter 2.

[16] CBO based carbon dioxide charges under the two hypothetical cap-and-trade policies on allowance prices in Sergey Paltsev and others, *Assessment of U.S. Cap-and-Trade Proposals*, Working Paper No. 13176 (Cambridge, Mass.: National Bureau of Economic Research, June 2007).

[17] For the most part, the assumptions that CBO adopted are drawn from analysis prepared by the Energy Information Administration.

[18] See John Deutch and others, *The Future of Nuclear Power: An Interdisciplinary MIT Study* (July 2003); and International Energy Agency, *World Energy Outlook* (2006).

[19] The calculation is based on data from Energy Information Administration, *An Analysis of Nuclear Power Plant Construction Costs*, DOE/EIA-0485 (1986). Those data include only plants on which construction was begun after 1965 and completed by 1986.

[20] See Pietro S. Nivola, "The Political Economy of Nuclear Energy in the United States," *Brookings Policy Brief No. 138* (September 2004).

[21] See Deutch and others, *The Future of Nuclear Power*, p. 142

[22] See David Gauthier-Villars, "Trials of Nuclear Rebuilding: Problems at Finland Reactor Highlight Global Expertise Shortage," *Wall Street Journal*, March 3, 2007, p. A6.

[23] Financing costs are also influenced by the period of capital recovery—the number of years over which the plant generates revenue for equityholders. As the recovery period increases, so do the financing costs.

[24] The 10 percent rate of return is based on 45 percent debt financing and 55 percent equity financing. Debt is assumed to be repaid at a rate of return of 8 percent over 20 years, and

equity is assumed to be repaid at an average rate of return of 14 percent over the 40 years the plant is assumed to operate.

[25] See "Coal Utilities Say They Do Not Fear Risk to Credit, Despite Moody's Warning on Carbon Burdens," *Platts Electric Utility Week* (March 3, 2008), p. 1. According to that report, Moody's Investors Services has warned that the prospect of future carbon dioxide charges may adversely affect the credit rating of utilities and thus raise the cost of capital for investment in conventional coal-fired generation.

[26] Fuel costs at innovative fossil-fuel plants are expected to be 10 percent to 30 percent higher because additional energy is needed to capture carbon dioxide.

[27] See Department of Energy, *Nuclear Waste Fund Fee Adequacy: An Assessment*, DOE/RW-0534 (May 2001), p. 1.

[28] See Joseph V. Spadaro, Lucille Langlois, and Bruce Hamilton, *Greenhouse Gas Emissions of Electricity Generation Chains: Assessing the Difference* (International Atomic Energy Agency, February 2000), p. 21.

[29] See Kathy Carolin Larsen, "Carbon Leads Long List of Electricity Market Risks," *Platts Insight* (November 2006).

[30] For more information, see Congressional Budget Office, *The Potential for Carbon Sequestration in the United States*.

[31] After canceling the original FutureGen program—which would have funded the construction of a single coal plant with CCS technology that also would have generated hydrogen for commercial purposes—the Department of Energy now plans to fund research and development for multiple innovative coal plants that use CCS technology.

[32] For a detailed review of the role of research and development in promoting technologies that reduce carbon dioxide emissions, see the Congressional Budget Office report, *Evaluating the Role of Prices and R&D in Reducing Carbon Dioxide Emissions* (September 2006).

[33] Although Nuclear Power 2010 and FutureGen have been two of the largest EPAct research and development incentives for the electricity industry, other cost-sharing programs exist for innovative coal technologies, renewable energy technologies, and fourth- generation nuclear reactors.

[34] The EPAct expansion of the preferential tax treatment of decommissioning funds reduces the private cost of cleaning and securing a nuclear facility once it is retired, which primarily involves the disposal of low-level radioactive waste. The federal government also plays a role in the long-term disposal of spent fuel, but that program is not addressed by EPAct.

[35] The production tax credit is also available to investors in some zero-emissions renewable technologies.

[36] Levelized costs for CCS technologies are not reflected in Figure 3-1 because those estimates were not available in all of the reports. In particular, the MIT study group did not analyze CCS technologies in *The Future of Nuclear Power* (2003). That study group estimated the cost of coal-fired power plants with CCS technology in a later report, *The Future of Coal* (2007), but changes in the methodology obscure whether the levelized cost for CCS technology in that report can be compared with the levelized cost of nuclear power in the earlier analysis. EIA found that the cost of generating electricity from coal-fired power plants with CCS would exceed the cost of power generated by nuclear power plants by about 15 percent.

[37] Marc Chupka and Gregory Basheda, *Rising Utility Construction Costs: Sources and Impacts* (report submitted by the Brattle Group to The Edison Foundation, September 2007), available at www.edisonfoundation.net/ Rising_Utility_Construction_Costs.pdf.

[38] "Samurai-Sword Maker's Reactor Monopoly May Cool Nuclear Revial," *Bloomberg.com*, at http://bloomberg.com/apps/ news ?pid=20670001 &refer=home&sid+aaVMzCTMz3ms.

[39] See *Budget of the United States Government, Fiscal Year 2009— Appendix*, p. 407, available at www.whitehouse.gov/omb/budget/ fy2009/.

[40] Those costs would include fixed payments for debt.

[41] See Members of the Special Committee on Nuclear Power, Wisconsin Legislative Council, Staff Memorandum No. 2 (November 2006), available at www.legis.state.wi.us/lc/committees/study/2006/NPOWR/files/memo2_npowr.pdf

[42] That increase is based on the Price Index for Fixed Investment in Power and Communication, produced by the Bureau of Economic Analysis. The data are adjusted for inflation using the price index for gross domestic product.

[43] The more stringent cap would reduce future emissions to 80 percent below 1990 emissions by 2050. CBO relates that cap to 2008 emissions for comparability with the less stringent cap.

In: Rethinking Nuclear Power in the United States ISBN: 978-1-60692-152-4
Editor: Kenneth A. Vellis

Chapter 2

NUCLEAR POWER: OUTLOOK FOR NEW U.S. REACTORS*

Larry Parker and Mark Holt

SUMMARY

Nearly three decades after the most recent order was placed for a new nuclear power plant in the United States, several utilities are now expressing interest in building a total of up to 30 new reactors. The renewed interest in nuclear power has resulted primarily from higher prices for natural gas, improved operation of existing reactors, and uncertainty about future restrictions on coal emissions. A substantial tax credit and other incentives for nuclear generation provided by the Energy Policy Act of 2005 (P.L. 109-58) are also likely to improve the economic viability of qualifying new reactors. New nuclear plant applications can also take advantage of amendments to the Atomic Energy Act made in the early 1990s to reduce licensing delays.

Currently, there are 103 licensed and operable power reactors at 65 plant sites in 31 states, generating about one-fifth of U.S. electricity. Although no new U.S. reactors have started up since 1996, U.S. nuclear electricity generation has since grown by more than 20%. Much of this additional output

* This is an edited, reformatted and augmented version of a CRS Report for Congress publication dated March 2007.

resulted from reduced downtime, notably through shorter refueling outages. Licensed commercial reactors generated electricity at an average of 89.8% of their total capacity in 2006, after averaging about 75% in the mid-1990s and about 65% in the mid-1980s.

Falling operating costs have helped renew the economic viability of the nation's fleet of nuclear power plants. From 1989 to 1998, 12 commercial reactors were closed before reaching the end of their 40-year licenses. By the late 1990s, there was real doubt that any reactors would make it to 40 years. Since 2000, however, 44 commercial reactors have received 20-year license extensions from the Nuclear Regulatory Commission (NRC), giving them up to 60 years of operation, and more are pending.

The nuclear production tax credit in the Energy Policy Act could have a significant impact on the economic viability of new nuclear power plants. Under base case assumptions, new reactors are not competitive with either coal-fired or natural gas-fired facilities. However, if new reactors are able to take full advantage of the nuclear production tax credit, nuclear power appears competitive with either natural gas-fired or coal-fired facilities.

Other factors will also be important in the commercial decision to invest in new nuclear plants, such as fossil fuel prices and the regulatory environment for both nuclear power and future fossil fuel-fired generation. If natural gas prices remain at historically high levels, future nuclear plants will be more likely to be competitive without federal tax credits. However, natural gas prices have been highly cyclical in the past, raising the possibility that nuclear costs could be undercut in the future.

Any substantial mandatory greenhouse gas control program would probably affect the cost of new coal-fired and natural gas-fired generation relative to nuclear power, particularly if nuclear power is assumed to have no greenhouse gas emissions. Continued delays in nuclear waste disposal facilities — forcing spent fuel to be stored at plant sites — could also affect the decision to construct new reactors.

Introduction

Construction of new nuclear power plants in the United States was almost unimaginable during the 1980s and 1990s. Vague rumors about possible new reactors would occasionally prompt a flurry of speculation, but they were invariably unfounded. In fact, no reactor has been ordered in the United States since 1978, and that plant was later cancelled, as eventually were all U.S. reactor orders after 1973. No U.S. reactor has been completed since 1996 — the Tennessee Valley Authority's Watts Bar 1, which had been ordered in 1970.

Today, there are still no orders, but interest in new U.S. reactors is no longer merely a rumor. In 2003, three utilities submitted applications to the Nuclear Regulatory Commission (NRC) for early approval of potential reactor sites under a cost-shared program with the Department of Energy (DOE). In 2004, DOE announced cost-sharing agreements with two industry consortia to apply for NRC licenses to construct and operate new reactors. Since then, a dozen more utilities and other companies have announced plans to apply for reactor licenses (as shown in **Table 1**), for a total of 34 new nuclear units. Several other companies have announced that they are considering filing applications as well.

The renewed interest in nuclear power has resulted primarily from higher prices for natural gas, improved operation of existing reactors, and uncertainty about future restrictions on coal emissions. Until the recent price volatility, low fuel costs had helped gas-fired power plants dominate the market for new electric generation capacity since the late 1980s. Nuclear power's relatively stable costs and low air emissions may now appear more attractive, particularly combined with a substantial tax credit for nuclear generation and other incentives provided by the Energy Policy Act of 2005 (P.L. 109-58). New nuclear plant applications can also take advantage of amendments to the Atomic EnergyAct made in the early 1990s to reduce licensing delays.[1]

In announcing the new reactor license applications, however, utilities have made clear that they are not committed to actuallybuilding the reactors, even if the licenses are approved. Large uncertainties about nuclear plant construction costs still remain, along with doubts about progress on nuclear waste disposal and concerns about public opposition. All those problems helped cause the long cessation of U.S. reactor orders and will need to be addressed before financing for new multibillion-dollar nuclear power plants is likely to be obtained.

Federal energy policy may play a crucial role in determining whether the current interest in new nuclear reactors leads to a significant expansion of the U.S. nuclear power industry. Nuclear opponents have long maintained that nuclear power will never be economically viable without federal subsidies and should be abandoned in favor of safer alternatives. But supporters contend that nuclear power will be vital in diversifying the nation's future energy supply and reducing greenhouse gas emissions, and that federal subsidies for at least the first few new reactors are justified. The greenhouse gas issue has also prompted some environmentalists to support nuclear power expansion.

Table 1. Announced Nuclear Plant License Applications

Announced Applicant	Site	Planned Application Date	Reactor Type	Units
Amarillo Power	Not specified	2007	GE ABWR	2
Constellation Energy(Unistar)	Calvert Cliffs (MD)	4Q 2007	Areva EPR	1
	Nine Mile Point (NY)	1st half 2008	Areva EPR	1
	Not specified	4Q 2008	Areva EPR	3
Dominion	North Anna (VA)	Nov. 2007	GE ESBWR	1
DTE Energy	Fermi (MI)	4Q 2008	Not specified	1
Duke Power	Cherokee (SC)	2007-2008	West. AP1000	2
Entergy	River Bend (LA)	May 2008	GE ESBWR	1
Exelon	Texas	Nov. 2008	Not specified	2
FPL	Not specified	2009	Not specified	1
NRG Energy	South Texas Project	2007	GE ABWR	2
NuStart	Grand Gulf (MS)	Nov. 2007	GE ESBWR	1
	Bellefonte (AL)	Oct. 2007	West. AP1000	2
Progress Energy	Harris (NC)	Oct. 2007	West. AP1000	2
	Levy County (FL)	July 2008	West. AP1000	2
SCE&G	Summer (SC)	3Q 2007	West. AP1000	2
Southern	Vogtle (GA)	Mar. 2008	West. AP1000	2
TXU	Comanche Peak (TX)	4Q 2008	Not specified	2
	Texas	4Q 2008	Not specified	2
	Texas	4Q 2008	Not specified	2
Total Units				**34**

Sources: NRC, *Nucleonics Week*, *Nuclear News*, Nuclear Energy Institute, company news releases.

This report includes analyses of the potential effect of the tax credit for nuclear power provided by the Energy Policy Act of 2005 and possible competitive effects of various proposals to limit greenhouse gas emissions. Under baseline assumptions, the cost of electricity from new nuclear power plants is likely to be higher than power generated by new coal- and natural gas-fired plants. The new nuclear tax credit would more than offset that cost disadvantage, but it is limited to the first 6,000 megawatts of new nuclear generating capacity. That is the capacity of about four to six reactors, although the credits could be spread among a larger number of new reactors under current rules. If the tax credit results in new reactor construction, the next question will be whether nuclear construction would continue without further credits. Greenhouse gas legislation could also be an important factor in nuclear power economics; analysis shows that some proposals, if enacted, could push the cost of coal- and natural gas-fired electricity above projected nuclear costs.

CURRENT STATUS OF U.S. NUCLEAR INDUSTRY

After the apparently successful commercialization of nuclear power in the 1960s, the Atomic Energy Commission anticipated that more than 1,000 reactors would be operating in the United States by the year 2000.[2] But by the end of the 1970s, it had become clear that nuclear power would not grow nearly that dramatically, and more than 120 reactor orders were ultimately cancelled. Currently, 103 licensed power reactors operate at 65 plant sites in 31 states (not including the Tennessee Valley Authority's [TVA's] Browns Ferry 1, which has not operated since 1985; TVA is spending about $1.8 billion to restart the reactor by mid-2007).

Despite falling short of those early expectations, however, U.S. nuclear power production has grown steadily since its inception and now exceeds electricity generated from oil, natural gas, and hydro plants, and trails only coal, which accounts for more than half of U.S. electricity generation. Nuclear plants generate more than half the electricity in six states. The near-record 823 billion kilowatt-hours of nuclear electricity generated in the United States during 2006[3] was more than the nation's entire electrical output in the early 1960s, when the first large-scale commercial reactors were being ordered, and more than twice the 2005 total electrical generation of Great Britain.[4]

As indicated in **Figure 1**, although no new U.S. reactors have started up since 1996, U.S. nuclear electricity generation has since grown by more than 20%.[5] Much of this additional output resulted from reduced downtime, notably through shorter refueling outages, which typically take place every 18 months.

Licensed commercial reactors generated electricity at an average of 89.8% of their total capacity in 2006, after averaging around 75% in the mid-1990s and around 65% in the mid-1980s.[6] Reactor modifications to boost capacity have also been a factor in the continued growth of nuclear power production. Since 1996, NRC has approved more than 60 requests for power uprates, totaling about 2,500 megawatts of electrical generating capacity — about the capacity of two large reactors.[7] The uprates largely offset the closure of five poorly performing reactors, totaling 3,700 megawatts of capacity, in 1997 and 1998.[8] Further uprate requests are pending.

The improved operation of nuclear power plants has helped drive down the cost of nuclear-generated electricity. Average operations and maintenance costs (including fuel but excluding capital costs) dropped steadily from a high of about 3.5 cents/kilowatt-hour (kwh) in 1987 to below 2 cents/kwh in 2001 (in 2001 dollars).[9] By 2005, the average operating cost was 1.7 cents/kwh.[10]

Falling operating costs haveimproved the outlook for the nation's existing fleet of nuclear power plants. From 1989 to 1998, 12 commercial reactors were closed before reaching the end of their 40-year licenses — California's Rancho Seco plant and Oregon's Trojan plant after only 14 and 16 years of operation, respectively.[11] By the late 1990s, there was real doubt about whether any reactors would make it to 40 years. Since 2000, however, 48 commercial reactors have received 20-year license extensions from NRC, giving them up to 60 years of operation. License extensions for seven more reactors are currently under review, and many others are anticipated, according to NRC.[12] The license extension trend has been spurred partly by favorable rate treatment of nuclear plants' unrecovered capital costs ("stranded costs") in states that have deregulated the power generation sector.

Industry consolidation could also help existing nuclear power plants, as larger nuclear operators purchase plants from utilities that run only one or two reactors. Several such sales have occurred, including the March 2001 sale of the Millstone plant in Connecticut to Dominion Energy for a record $1.28 billion. The merger of two of the nation's largest nuclear utilities, PECO Energy and Unicom, completed in October 2000, consolidated the operation of 17 reactors under a single corporate entity, Exelon Corporation, headquartered in Chicago.

Although no new U.S. nuclear power plant has opened in the past 10 years, commercial reactor construction has continued elsewhere in the world, particularly in Asia. Since the most recent U.S. reactor began operating in 1996, 37 have started up in other countries, an average of about four per year.[13] Twenty-five reactors are currently under construction outside the United States.[14]

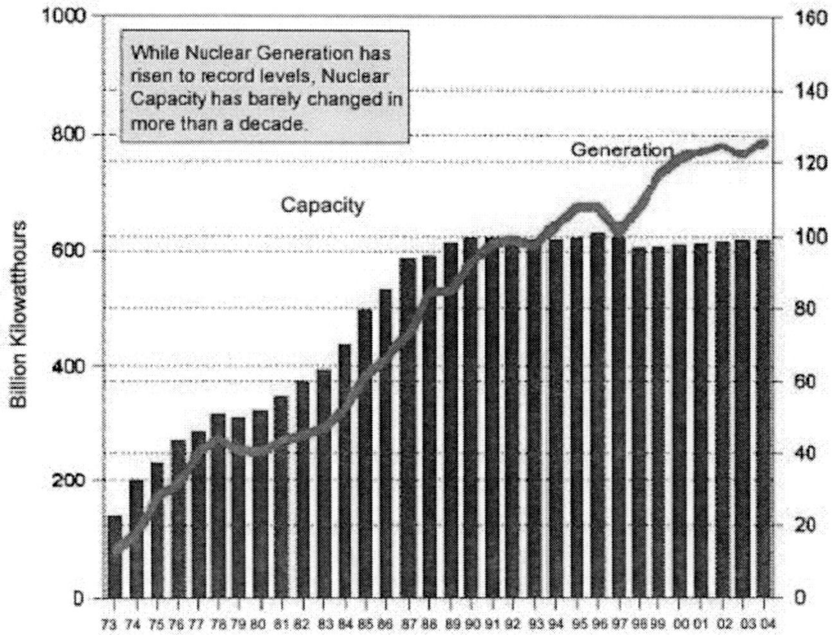

Source: Energy Information Administration.

Note: Generation is read on the left scale (in billion kilowatt-hours) and capacity (in gigawatts) is on the right.

Figure 1. Net Nuclear Generation vs. Capacity, 1973-2004

FEDERAL INITIATIVES TO ENCOURAGE NEW NUCLEAR POWER PLANT CONSTRUCTION

With the Energy Policy Act of 2005, the federal government has adopted aggressive incentives for building new reactors — including tax credits, loan guarantees, and compensation for regulatory delays. These incentives build on previous regulatory and legislative initiatives, particularly a more streamlined NRC licensing process and DOE's Nuclear Power 2010 program to test that process.

NRC Licensing Reform

Until 1989, licensing a new nuclear power facility involved a two-step process: (1) an NRC-issued construction permit that allowed an applicant to begin building a facility and (2) an operating license that permitted the facility to generate electricity for sale.[15] This procedure resulted in some celebrated cases in which completed or nearly completed plants awaited years to be granted operating licenses — delays that drove up the costs of the affected plants. In 1989, NRC issued regulations to streamline this process in three ways:[16]

- The Early Site Permit Program allows utilities to get their proposed reactor sites approved by the NRC before a decision is made on whether or not to build the plant. These preapproved sites can be "banked" for future use.
- Standard Design Certification for advanced reactor designs allows vendors to get their designs approved by NRC for use in the United States, so utilities can then deploy them essentially "off the shelf."
- The Combined Construction and Operating License (COL) provides a "one-step" approval process, in which all licensing hearings for a proposed plant are expected to be conducted before construction begins. The COL would then allow a completed plant to operate if inspections, tests, analyses, and acceptance criteria (ITAAC) were met. This is intended to reduce the chances for regulatory delays after a plant is completed.[17]

The relationships among these three components are illustrated in **Figure 2**. A COL application could reference a preapproved site and a certified plant design, so that most siting and design issues would not need to be revisited. Upon completion, the reactor could begin operating after NRC determined that the ITAAC had been met. The overall goal of these reforms was to introduce as much regulatory certainty into the process before a company has to make a major financial investment in a project. However, the process has never been used, so it remains uncertain how much time will be saved by referencing preapproved sites and certified designs, or how difficult the ITAAC checkoff process might be.

The procedures envision a three-step decision-making process, allowing the utility to make "go/no-go" decisions at several points before a major investment is made in the project. The first step to building a new facility is to conduct utility level project analysis, including needs assessment, environmental impact scoping analysis, and identification of siting issues. This is anticipated to take about 2-4 years, and some utilities have already begun this process (see **Table 1**). Assuming the utility finds nuclear power to be a viable option, it will have to address three issues in this pre-application process. First, the utility will have to evaluate safety-related issues, such as seismic and geologic data, population demographics, and potential consequences of hypothetical accidents. Second, the utility will have to evaluate environmental issues, such as maximum radiological and thermal effluents. Third, the utility will have to address emergency planning issues, such as evacuation routes. The utility may use the NRC early site permit process to conduct this evaluation but is not required to do so.

Assuming the utility finds the site suitable and the project potentially economical, it submits the above information, along with further details, to the NRC to obtain a COL.This additional information includes financial data, justification for the capacity addition, and complete details on reactor design. On this last point, the utility is likely to reference a standard design certification but is not required to do so. The utility must also provide the ITAAC for the eventual NRC approval to operate the plant. The NRC reviews the application, holds hearings, and makes a decision on granting the COL.

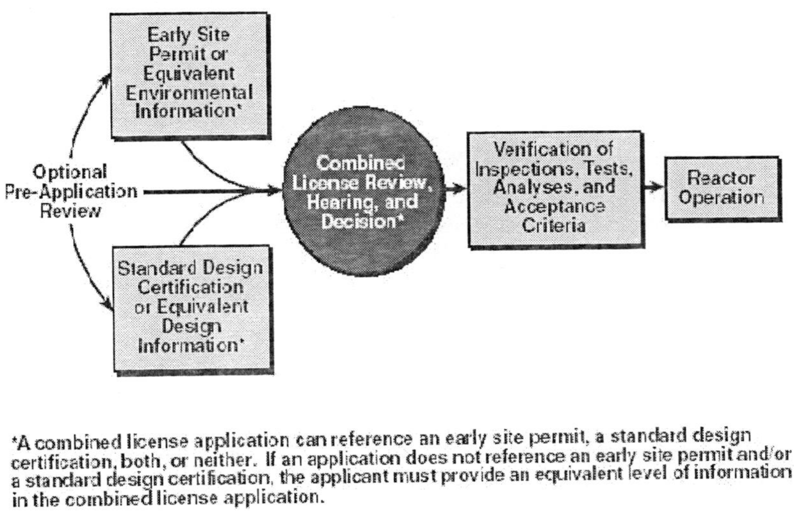

Source: NRC.

Figure 2. Relationship between Combined Licenses, Early Site Permits, and Standard Design Certifications

The licensing process is currently estimated by NRC to take about three and a half years, although NRC Chairman Dale Klein has called for that schedule to be shortened.[18] NRC established the Office of New Reactors in 2007 to handle the potential influx of new reactor license applications.

After the license is issued, the utility must decide whether to begin building the power plant. Current projections of nuclear power construction schedules assume that a plant can be built in 5-7 years. At the end of construction, the NRC verifies thatthenew plant meets the ITAAC in the COL and the facility is allowed to operate. Overall, the process is anticipated to take 10-15 years.

If this streamlined process works as intended, it may remove some of the previous regulatory uncertainty surrounding new nuclear plant construction and make financing of such projects more feasible. This is particularly true for the roughly half of the states that have restructured their electricity markets, thus resulting in utilities employing project financing rather than more traditional funding. With project financing, the proposed developer of a power plant seeks financing for the project using only the project as recourse for the loan, as opposed to securing the loan with the larger holdings of the utility itself. With the project being the only collateral, Wall Street looks very closelyat the risk profile of the project in determining whether to finance it and on what terms. The nuclear industry and the NRC hope that the new licensing process will help improve the risk profile of new facilities by increasing the certainty that a plant will be built expeditiously and begin operations in a timely manner. It is also possible that an increase in nuclear power plant permit applications could make the new process more routine, shortening approval time (as has happened with licensing renewal requests for existing facilities, which are now generally approved in about 18 months).

However, there are several reasons to believe that the longer end of the 10-15 year range is more likely, at least in the short-term. First, this is an untried process, as noted above. Uncertainties include some time-honored ones, such as the environmental impact statement and safety evaluation report, as well as new issues presented by the new procedures, such as NRC's certification of a utility's ITAAC. Second, public input is likely to be vigorous. Initial efforts by utilities to obtain early site permits have been slowed by substantial public comments on each permit request. Third, the new procedures do not prevent state intervention into the process, particularly where traditional rate-making authority remains. States can be very resourceful in delaying nuclear power when they so choose. The classic example is the Shoreham nuclear power station, which was completed and licensed but never began commercial operation because of the sustained opposition from the state of New York.

Finally, judicial intervention is not unusual when opposition interest groups attack permits, environmental impact statements, and other regulatory decisions in attempts to forestall construction and operation. When existing nuclear plants were licensed, opposition often focused on the potential for reactor accidents. In the post-9/11 environment, concerns about terrorist attacks are likely to be raised as well.[19]

DOE Nuclear Power 2010 Program

Because no early site permits or COLs had ever been sought, DOE in 2002 initiated the Nuclear Power 2010 Program to demonstrate those processes, offering to pay up to half the licensing costs incurred by industry applicants. The program's original goal was to pave the way for deployment of at least one new nuclear power plant by 2010 and thus reduce regulatory uncertainty for further license applicants.

Although the program's original goal of deploying a reactor by 2010 will not be achieved, industry interest in the effort has been substantial. Under the program, three utilities applied to NRC in 2003 for early site permits to build new reactors at existing plants in Illinois, Mississippi, and Virginia. NRC approved the Illinois permit March 8, 2007, and anticipates final action on the others later this year.[20]

Two industry consortia will continue to receive DOE assistance over the next several years to apply for COLs and conduct "first of a kind engineering" for new nuclear power plants, although they have not committed to ordering the reactors if the licenses are issued. DOE awarded the first funding to the consortia in 2004. DOE assistance under the program, including the early site permits, is planned to reach a multiyear total of about $550 million. The two consortia receiving COL assistance under the Nuclear Power 2010 program are

- A consortium led by Dominion Resources that is preparing a COL for an advanced General Electric reactor (after originally considering a Canadian design). The proposed reactor would be located at Dominion's existing North Anna plant in Virginia, where the company is seeking an NRC early site permit with DOE assistance.

- A consortium called NuStart Energy Development, which includes Exelon and several other major nuclear utilities. The consortium announced on September 22, 2005, that it would seek a COL for a Westinghouse design at the site of TVA's uncompleted Bellefonte nuclear plant in Alabama and for a General Electric design at the Grand Gulf plant in Mississippi.

Energy Policy Act of 2005

Nuclear Production Tax Credit

The most direct nuclear incentive provided by the Energy Policy Act is a 1.8 cents/kwh tax credit for up to 6,000 megawatts of new nuclear capacity for the first eight years of operation, up to $125 million annually per 1,000 megawatts. An eligible reactor must be placed into service before January 1, 2021. As discussed below, this credit is expected to significantly improve the projected economic viability of proposed nuclear power plants.

A major factor in determining the potential impact of the nuclear production credit is the allocation of the credit among eligible reactors. Under the Energy Policy Act, the 6,000 megawatts of capacity that could receive the credit is to be allocated by the Secretary of the Treasury in consultation with the Secretary of Energy. The Internal Revenue Service issued interim guidance on May 1, 2006, that would provide the tax credit to electricity generated by any reactor that (1) applied for an NRC combined license by December 31, 2008; (2) began construction before January 1, 2014; and (3) was certified by DOE as meeting eligibility requirements.[21]

Under the guidance, if license applications with more than 6,000 megawatts of eligible nuclear capacity are received by December 31, 2008, the 6,000-megawatt cap will be allocated proportionally among the eligible plants. Therefore, if 12,000 megawatts of new nuclear capacity met the application deadline and eventually went into operation, then only half the electrical output of each reactor would get the tax credit. If license applications by December 31, 2008, totaled less than 6,000 megawatts, then additional reactors would become eligible until the limit is reached, according to the IRS guidance.

The deadline for automatic eligibility for the tax credit appears to provide a strong incentive for nuclear plant applicants to file with NRC by the end of 2008, which is sooner than some of the anticipated filings shown in **Table 1**.

However, if most of those reactors were to become eligible for the credit, the credit's effect could be diluted to the point where it would no longer provide a sufficient construction incentive.

The credit would most dramatically affect nuclear plant economics if 100% of a reactor's output were eligible; however, if each new nuclear unit were to receive the credit for all its electrical generation, then only four or five reactors (ranging from 1,200-1,500 megawatts) could be covered within the 6,000-megawatt limit. Because reactor designs from three different companies are currently under consideration, only one or two units of each design might be constructed under this scenario. That might not be enough to reduce costs through series production to the point where further units — ineligible for tax credits — would be economically viable on their own.

Regulatory Risk Insurance

Continuing concern over potential regulatory delays, despite the streamlined licensing system now available, prompted Congress to include an insurance system in EPACT that would cover some of the costs of such delays. The regulatory delay insurance, called "Standby Support," would cover the principal and interest on debt and extra costs incurred in purchasing replacement power because of licensing delays. The first two new reactors licensed by NRC that meet other criteria established by DOE could be reimbursed for all such costs, up to $500 million apiece, whereas each of the next four newly licensed reactors could receive 50% reimbursement of up to $250 million.

DOE issued the final rule for the Standby Support program on August 11, 2006.[22] Criteria established by the rule for receiving coverage under the program include the issuance of a COL, a detailed construction schedule, documentation that construction has started, and a detailed schedule for completing the inspections, tests, analyses, and acceptance criteria required for reactor operation to begin. The first two reactors to meet all the criteria would receive the $500 million coverage.

Coverage is to be provided for delayscaused by NRC's failure to follow its own rules (if any) in reviewing a reactor's ITAAC, NRC's failure to meet DOE-approved ITAAC schedules, NRC pre-operational hearings, and litigation. Standby Support coverage is not provided for delays caused by "failure of the sponsor to take any action required by law, regulation, or ordinance."[23] This includes delays caused by NRC orders to re-conduct ITAAC or to correct pre-operational deficiencies found by NRC. However, the program does cover delays caused by licensing-related litigation in state, federal, or tribal courts, even if a court rules against a nuclear plant sponsor.[24]

The Standby Support program is intended to reduce uncertainty about the COL licensing process that may pose an obstacle to nuclear plant orders. Because the first two reactors would presumably face the most uncertainty about the untried process, they would receive the most coverage. It is apparently hoped that the licensing experience of the first two reactors would provide enough confidence for the next four to proceed with half the coverage, and then for additional reactors to be built with no regulatory risk insurance.

Loan Guarantees

New nuclear power plants are eligible for federal loan guarantees authorized by EPACT for energy projects that reduce air emissions, a criterion that includes new clean coal projects. The loan guarantees may cover up to 80% of a plant's estimated cost. If a borrower defaults, DOE is to pay off the loan and can either (1) take over the project for completion, operation, or disposition or (2) reach an agreement with the borrower to continue the project. To prevent default, DOE may make loan payments on behalf of the borrower, subject to appropriations and an agreement by the borrower for future reimbursement.

Because it is generally believed that Wall Street continues to view new commercial reactors as financially risky, the availability of federal loan guarantees could be a key element in attracting funding for such projects and reducing financing costs. The federal government would bear most of the risk, facing potentially large losses if borrowers defaulted on reactor projects that could not be salvaged. Loan guarantees may be especially important for nuclear projects undertaken by deregulated generating companies as opposed to traditionally regulated utilities, which can recover their regulator-approved capital costs from ratepayers. Even for regulated utilities, "loan guarantees are critically important to new nuclear plant financing," the Nuclear Energy Institute contended in September 2006 testimony.[25]

DOE issued its initial solicitation for loan guarantees under EPACT on August 8, 2006.[26] The total amount of the loan guarantees in the initial solicitation is limited to $2 billion and does not include nuclear technology. Along with the initial solicitation, DOE issued guidelines for considering those initial proposals and said it was working on final regulations to govern future loan guarantee solicitations. The deadline for applications under the initial solicitation was originally November 6, 2006, but was subsequently delayed to December 31, 2006.

EPACT requires that before a loan guarantee is granted, the estimated subsidy cost (including estimated default losses) must be covered by a specific appropriation or by an up-front payment from the borrower. DOE's initial solicitation says that, because appropriations for the program are not anticipated, each borrower will have to pay the estimated subsidy cost. According to the solicitation, the subsidy cost will be calculated for each loan and must be approved by the Office of Management and Budget (OMB).[27] OMB will undoubtedly want to ensure that the payments are high enough to cover all the anticipated default and other subsidy costs incurred by the DOE loan guarantee program. The size of the payments required by OMB could strongly affect the value of the loan guarantees to borrowers.

Also important to potential borrowers is the percentage of project costs that can be covered by the DOE loan guarantees. Although EPACT Section 1702(c) allows DOE to provide loan guarantees for up to 80% of a project's estimated cost, DOE's guidelines for the initial solicitation "expresses a preference" that the loan guarantees cover not more than 80% of a project's debt.[28] Therefore, if a project has significant non-debt financing, the loan guarantees could cover considerably less than 80% of the total cost. The Nuclear Energy Institute contended that "the procedures outlined in the guidelines are so restrictive and so conditional that they would not support financing of a nuclear power plant."[29]

Analysis of New Nuclear Power Plant Construction

Base Case Assumptions

To examine the potential competitive position of new nuclear plants, the future market price for electricity must be estimated. In the long run, the marginal cost of bringing on new electricgeneratingcapacitywill tend to be set by the cost of the least expensive newly constructed generating plant. To evaluate the potential competitive position of nuclear power, CRS constructed an illustrative example involving four hypothetical powerplants: a conventional pulverized coal-fired facility, an advanced coal-fired facility based on integrated gasification combined-cycle (IGCC) technology, an advanced natural gas-fired combined-cycle facility, and an advanced nuclear power facility. The illustrative examples permit a consistent set of assumptions to use for the analysis.

Assessing the competitiveness of future nuclear power plants requires numerous assumptions about future economic, financial, and policy conditions. Based on 2015 as the benchmark year for constructing a new power plant, the major assumptions of the analysis are identified in **Table 2**. Most of the assumptions are from the Energy Information Administration (EIA) and reflect costs and technical performance anticipated by EIA for projects initiated in 2015.[30] The real capital charge rate is from the Environmental Protection Agency's Integrated Planning Model (IPM).[31] Calculations were done by CRS and are in constant 2004 dollars.

For nuclear fuel and coal costs, the assumptions reflect the price trends projected by EIA during the construction period. For coal, EIA projects stable prices in real terms from 2015 through 2023. For uranium, EIA projects stable prices in real terms between 2015 and 2030. Because natural gas prices have historically been more volatile than coal prices, CRS has performed a sensitivity analysis on natural-gas generating costs using a range of potential future natural gas prices. This sensitivity analysis, along with a general discussion of natural gas and coal prices, is presented after the base case results, which assume stable (in real terms) projected 2015 prices.

Table 2. Projected 2015 Costs and Assumptions (2004$)

Assumption	Coal Plant	Advanced Coal Plant	Advanced Natural Gas GCC	Advanced Nuclear Power Plant
Capital costs	$1,217/kw	$1,386/kw	$555/kw	$1,913/kw
Construction schedule	4 years	4 years	3 years	6 years
Fixed O&M costs	$25.07/kw-year	$35.21/kw-year	$10.65/kw-year	$61.82/kw-year
Variable O&M costs	0.418 cents/kwh	0.265 cents/kwh	0.182 cents/kwh	0.045 cents/kwh
Fuel costs	$1.40/million Btu	$1.40/ million Btu	$5.08/million Btu	$0.66/ million Btu
Heat rate	8,661 Btu/kwh	7,477 Btu/kwh	6,403 Btu/kwh	10,400 Btu/kwh
Capacity factor	90%	90%	90%	90%
Real capital charge rate	13.4%	13.4%	13.4%	13.4%
Date project initiated	2015	2015	2015	2015

Sources: DOE/EIA, *Assumptions to the Annual Energy Outlook 2006* (March 2006); EPA, *Standalone Documentation for EPA Base Case 2004* (September 2005).

Table 3. Projected 2015 Annualized Costs (2004$)

	Pulverized Coal Plant	Advanced Coal Plant	Advanced Natural Gas GCC	Advanced Nuclear Power Plant
Cents per kwh	4.5	4.6	4.6	5.6

Source: CRS calculations based on **Table 2** assumptions.

Base Case Reults

Assuming that by 2015 the subsidies contained in the 2005 Energy Policy Act are no longer available for new nuclear power construction, **Table 3** indicates that under EIA's assumed 2015 natural gas price scenario, conventional coal-fired and advanced combined-cycle natural gas-fired facilities would be in a virtual dead-heat as the choice for new construction. CRS estimates the annual costs on a levelized basis for new coal-fired or natural gas-fired facilities to be within one mill per kilowatt-hour (kwh) under EIA's estimated 2015 natural gas prices. Advanced coal-fired technology is projected to be competitive with both pulverized coal combustion and natural gas combined-cycle technology by 2015.

Without the production tax credit contained in the 2005 Energy Policy Act, a nuclear facility is not competitive with either coal-fired or natural gas-fired facilities under base case assumptions. Based on the assumptions above, CRS estimates that the break-even point for nuclear power capital costs versus coal-fired facilities initiated in 2015 is about $1,370 per kilowatt (kw) of capacity. This is substantially below the EIA projected cost of $1,913 per kw and is even below the vendors' estimate of $1,528 per kw (2004).[32] Under base case conditions, it seems unlikely that a new nuclear power plant would be constructed in the United States, barring a sustained, long-term increase in natural gas prices and the creation of a substantial, mandatory greenhouse gas reduction program that would increase coal-fired and natural gas-fired generating costs.

Table 4. Projected 2015 Annualized Costs, Including Subsidized Nuclear Power (2004$)

	Pulverized Coal Plant	Advanced Coal Plant	Advanced Natural Gas GCC	Advanced Nuclear Power Plant	Advanced Nuclear Power Plant with Production Credit
Cents per kwh	4.5	4.6	4.6	5.6	4.2-4.7[a]

Source: CRS calculations based on **Table 2** assumptions.

a. Range reflects uncertainty with future inflation and construction times and dates, which affect the real value of the production credit.

Impact of 2005 Energy Policy Act

However, if one assumes that the production tax credit contained in the 2005 Energy Policy Act is available to facilities that begin construction in 2015 (the base year in the analysis), the story is different. As indicated in **Table 4,** the production tax credit contained in EPACT is sufficient to make nuclear power competitive with either natural gas-fired or coal-fired facilities.

The advantage provided to nuclear power by the production tax credit is not definitive; however, it appears sufficient to allow a decision on constructing a nuclear power station to move beyond initial economic considerations to examining other relevant factors, such as fossil fuel prices and the regulatory environment for both nuclear power and future fossil fuel-fired generation.

Sensitivity Analysis

Volatile Natural Gas Prices

Relatively high natural gas prices and the country's two-decade reliance on natural gas for new electric generating capacity have raised concern that the country is becoming too dependent on natural gas for electricity. Currently, about 22% of the country's electric generating capacity is natural gas-fired, compared with about 7% two decades ago. This situation raises at least two questions: (1) whether the recent rise in natural gas prices is a harbinger of future prices or just another peak in the historic boom-bust cycle of U.S. natural gas prices, and (2) whether nuclear power is an alternative generating option that the federal government should subsidize to help address question number one.

The potential for increased natural gas prices has been illustrated by recent events. However, power plants are long-lived facilities with lifetimes estimated at around 65 years, and current maintenance practices can extend that life almost indefinitely. In the emerging competitive environment for new power plant construction, the financial investment lifetime may be on the order of 20 years. This shorter time frame to recover a power plant investment reflects the uncertainty existing in the electricity market.[33]

Twenty years is a long time in the natural gas and coal markets. **Figure 3** charts natural gas prices to electric utilities during the 38 years from 1967 to 2005. Converted to real 2000 dollars using the Implicit Price Deflator, the chart indicates that natural gas prices have generally stayed under $4.50 per thousand cubic feet (mcf) until 2003. It is uncertain how long the current relatively high prices may continue. As indicated in **Table 2**, EIA projects 2015 natural gas prices at $5.08 in 2004 dollars — above the levels of the 1990s, but below the prices of the last two years.

To illustrate the sensitivity of natural gas-fired electricgeneration to natural gas prices, **Table 5** provides estimates of generation costs for a range of natural gas prices. Based on these calculations, the breakeven point for unsubsidized nuclear power versus natural gas-fired facilities would be $6.65/mcf or million Btu (MMBtu) in 2004$. Thus, if the current level of natural gas prices continues in the long-term, nuclear power may not need any subsidies by the year 2015 to compete with natural gas-fired facilities.

Table 5. Effect of Natural Gas Prices on Production Costs (2004$)

Natural Gas Price (delivered, $/MMBtu)	Production Costs (cents/kwh)
3.08	3.3
4.08	4.0
5.08	4.6
6.08	5.2
7.08	5.9

Source: CRS calculations based on **Table 2** assumptions.

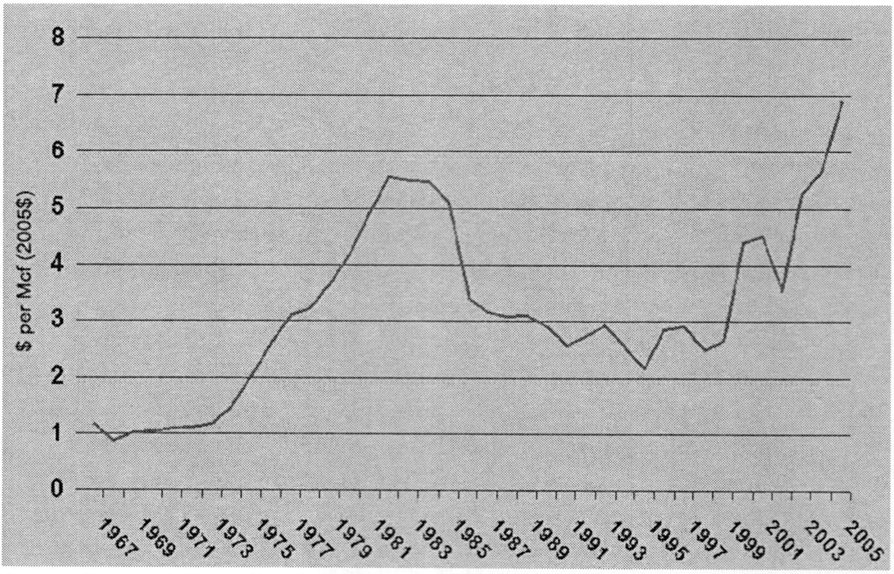

Source: Energy Information Administration, data available at [http://www.eia.doe.gov/emeu/aer/txt/stb0608.xls].

Figure 3. Natural Gas Prices Delivered to Electric Utilities

As noted above, the economic question raised by this analysis is whether the current upturn in natural gas prices is a relatively short-term phenomenon, or does it reflect a new long-term premium for natural gas, based on its environmental and technological advantages and future availability? If the former, then building new nuclear power plants would be a questionable venture economically, unless their capital costs could be reduced substantially. If the latter, nuclear power construction could become attractive in the future if prices persist at above approximately $6.65 per MMBtu (2004$).

However, it should be noted that volatile natural gas prices do not have any direct effect on generating costs at coal-fired facilities. In contrast to natural gas prices, coal prices generally have been on a slow, steady decline for 30 years. The increasing share of coal being supplied by large, low-cost surface mining operations in the West has contributed to a long-term downward trend in coal prices. Coal prices have risen in the past couple of years as demand has increased; however, with abundant reserves, a sustained increase would seem problematic.

Greenhouse Gas Control

Any substantial mandatory greenhouse gas control program would probably affect the cost of new coal-fired and natural gas-fired generation. In all current proposals before the Congress, nuclear power is assumed to have no greenhouse gas emissions. This "green" nuclear power argument has gotten some traction in think tanks and academia. As stated by MIT in its major study *The Future of Nuclear Power*: "Our position is that the prospect of global climate change from greenhouse gas emissions and the adverse consequences that flow from these emissions is the principal justification for government support of the nuclear energy option."[34] The industry also has been attempting to promote nuclear power as one solution to rising greenhouse gas emissions.[35] A few well-known environmentalists have expressed public support for nuclear power as part of the response to global climate change, although no major environmental group as yet has publically adopted that position.[36]

Despite strong Bush Administration opposition to mandatory greenhouse gas reduction programs, a number of congressional proposals to advance programs designed to reduce greenhouse gases were introduced in the 109th Congress,[37] and similar efforts have continued in the 110th Congress. None of these proposal have passed either house of Congress. The first effort to pass a mandatory greenhouse gas reduction program failed in 2003 on a 43-55 vote in the Senate. A similar effort was defeated in 2005 during the debate on the Energy Policy Act of 2005 on a 38-60 vote. This second, less favorable vote reflects the changed votes of four Senators who reportedly objected to the addition of nuclear power incentives to the 2005 version of the proposed legislation.[38] The proposals would have placed a cap on U.S. greenhouse gas emissions based on a 2001 baseline. The cap would have been implemented through a tradeable permit program to encourage efficient reductions.

However, concern that global climate change should be addressed by the Congress led 13 Senators to introduce S.Amdt. 866 — a Sense of the Senate resolution on climate change — during the debate on the Energy Policy Act of 2005.

The resolution finds that (1) greenhouse gases are accumulating in the atmosphere, increasing average temperatures; (2) there is a growing scientific consensus that human activity is a substantial cause of this accumulation; and (3) mandatory steps will be required to slow or stop the growth of greenhouse gas emissions. Based on these findings, the resolution states that it is the sense of the Senate that the Congress should enact a comprehensive and effective national program of mandatory, market-based limits and incentives on greenhouse gases that slow, stop, and reverse the growth of such emissions.

This should be done in a manner that will not significantly harm the U.S. economy and will encourage comparable action by other countries that are the nation's major trading partners and contributors to global emissions. The resolution passed by voice vote after a motion to table it failed on a 43-54 vote.

Four proposals introduced in the 109th Congress had analyses conducted by EPA and others that estimated their costs. For the 110th Congress, five bills have been introduced, but none have accompanying cost analyses.[39] However, several of the bills analyzed during the 109th Congress have been reintroduced in the 110th Congress, although with some modifications. These include S.317, which has the CO_2 control provisions of S. 2724 of the 109th Congress, S.280 and H.R. 620, which are modified versions of S. 1151 of the 109thCongress, and continuing efforts to draft legislation based on recommendations by the National Commission on Energy Policy (NCEP). The bills of the 109th Congress included the following:

- S. 2724 (Senator Carper).[40] Would create a cap-and-trade permit program to reduce emissions of sulfur dioxide, nitrogen oxides, mercury, and carbon dioxide from electric generating facilities greater than 25 megawatts (mw). The CO_2 cap would be set in two phases, with affected facilities required to reduce emissions to 2006 levels by 2010, and then further reduce emissions to 2001 levels by 2015.
- NCEP Recommendation (draft legislation prepared by Senator Bingaman).[41] Would create an economy-wide tradeable permit program to begin limiting greenhouse gases. The proposal would mandate an accelerated reduction in the country's greenhouse gas intensity: Between 2010 and 2019, the proposal would require a 2.4% annual reduction in greenhouse gas emissions per dollar of projected gross domestic product (GDP). After 2019, this reduction would increase to 2.8% annually. The program would include a cost-limiting safety valve that allows covered entities to make a payment to DOE in lieu of reducing emissions. The initial price of such payments would be $7 per ton in 2010, rising 5% annually thereafter.[42]
- S. 1151 (Senators McCain and Lieberman).[43] Would create an economy-wide cap-and-trade program to reduce emissions of six greenhouse gases to their 2000 levels by the year 2010. The flexible, market-based program would permit participation in pre-certified international trading systems and a carbon sequestration program to achieve part of the reduction requirement. The bill excludes

residential and agricultural sources, along with entities that do not own a single facility that emits more than 10,000 metric tons of CO_2 equivalent annually.[44]

- S. 150 (Senator Jeffords).[45] Would create a cap-and-trade permit program to reduce emissions of sulfur dioxide, nitrogen oxides, and carbon dioxide, along with unit-by-unit controls on mercury emissions from electric generating facilities 15 mw or greater. The CO_2 cap would require affected entities to reduce their emissions to 1990 levels by 2010.

Table 6 indicates projected trade permit prices for the four proposals. As indicated, S. 2724 is estimated to have the lowest price, while S. 150 is projected to have the greatest. This differential reflects both the stringency of the various proposals and their scope (economy-wide versus electric generation only). The reader should note that the estimates come from a variety of sources, and significant uncertainty surrounds the actual cost of any greenhouse gas initiative (except for the NCEP proposal, which includes a safety valve that limits the upper price range on permits).

Table 6. Per-Ton CO_2 Permit Price Estimates for Greenhouse Gas Initiatives (in 2004$/metric ton of CO_2)

Year	S. 2724[a]	NCEP Recommendations	S. 1151	S. 150
2015	$1.2	$5.9	$11.7	$25.8
2020	$2.5	$7.7	$14.9	$33.1

Sources: For S. 843 and S. 150: EPA, Office of Air And Radiation, *Multi-Pollutant Analysis: Comparison Briefing* (October 2005); for NCEP Recommendations: The National Commission on Energy Policy, *Ending the Energy Stalemate: A Bipartisan Strategy to Meet America's Energy Challenges* (December 2004); for S. 1151: Sergey Palsev, *et al.*, *Emissions Trading to Reduce Greenhouse Gas Emissions in the United States: The McCain Lieberman Proposal* [S. 139], Report No. 97 (June 2003). Estimates converted into 2004$ using the GNP Implicit Price Deflator.

a. S. 2724 estimates based on analysis of S. 843 introduced in 108th Congress. The later deadlines in S. 2724 would probably result in slightly lower cost estimates than those presented here, all else being equal.

Table 7 indicates the effect that the four proposals would have on 2015 and 2020 generation costs by fuel source. As indicated, the first two proposals, S. 2724 and the draft proposal based on the NCEP recommendations, would have a minimal effect on fuel choice in 2015 and 2020, all else being equal. The third bill, S. 1151, would pull nuclear power about even with its coal-fired competition. The fourth, S. 150, would provide the greatest advantage to nuclear power. It is also the proposal that most resembles the requirements of the Kyoto Protocol (as least for electric generation).

Table 7. 2015 and 2020 Projected Annualized Costs with Increased Costs from Greenhouse Gas Legislation (cents per kwh, 2004$)

	Pulverized Coal Plant		Advanced Coal Plant		Advanced Natural Gas GCC		Advanced Nuclear Power Plant
	2015	**2020**	**2015**	**2020**	**2015**	**2020**	
S. 2724[a]	4.6	4.7	4.7	4.8	4.6	4.7	5.6
NCEP Recommendations	4.9	5.1	5.0	5.1	4.8	4.9	5.6
S. 1151	5.4	5.7	5.4	5.6	5.0	5.1	5.6
S. 150	6.5	7.1	6.4	6.9	5.5	5.7	5.6

Source: CRS calculations based on **Table 6** estimates and **Table 2** assumptions.

a. S. 2724 estimates based on S. 843 introduced in 108th Congress. The later deadlines in S. 2724 would probably result in slightly lower cost estimates than those presented here, all else being equal.

To quantify the potential effect of permit prices (or an equivalent carbon tax) on fuel source for future electric generating capacity, **Table 8** provides the effects of a range of permit prices/carbon taxes on new electric generating cost under base case conditions. As indicated, the breakeven point for nuclear power versus natural gasfired facilities is about $30 a metric ton (2004$); the breakeven point for nuclearpower versus coal-fired facilities is about $15 a metric ton (2004$). Thus, over time, nuclear power could provide a form of safety valve for electric generation, if permit prices or a carbon tax became a permanent part of the electricity supply environment.

Nuclear Waste

Highly radioactive spent fuel produced by nuclear reactors poses a disposal problem that could be a significantfactor in the consideration of new nuclear plant construction. The Nuclear Waste Policy Act of 1982 (NWPA, P.L. 97-425) commits the federal government to providing for permanent disposal of spent fuel in return for a fee on nuclear power generation.

However, the schedule for opening the planned national nuclear waste repository at Yucca Mountain, Nevada, has slipped far past NWPA's deadline of January 31, 1998. DOE currently hopes to begin receiving waste at Yucca Mountain by 2017.[46]

Table 8. Effect of Permit Prices/Carbon Tax on Electricity Production Costs

Permit Price or CarbonTax (2004 \$/metric ton of CO_2)	Natural Gas (cents/kwh)	Conventional Coal (cents/kwh)	Advanced Coal (cents/kwh)
\$5	4.8	4.9	5.0
\$10	4.9	5.3	5.3
\$15	5.1	5.7	5.7
\$20	5.3	6.1	6.0
\$25	5.4	6.5	6.4
\$30	5.6	6.9	6.7
\$35	5.8	7.3	7.0
\$40	6.0	7.7	7.4

Source: CRS calculations based on **Table 2** assumptions.

In the meantime, more than 50,000 metric tons of spent fuel is being stored in pools of water or shielded casks at nuclear facility sites.[47] NWPA limits the planned Yucca Mountain repository to the equivalent of 70,000 metric tons of spent fuel. Because U.S. nuclear power plants discharge an average of 2,000 metric tons of spent fuel per year, the Yucca Mountain limit is likely to be reached before any new reactors begin coming on line.

Therefore, even if Yucca Mountain eventually begins operating as planned, it is unclear what ultimately would be done with spent fuel from new nuclear power plants under current law. In the near term, continued storage at reactor sites and interim storage at central locations would be the most likely possibilities. The primary long-term options include lifting the statutory cap on Yucca Mountain disposal, developing additional repositories, and reprocessing spent fuel for reuse of plutonium and uranium. The Bush Administration's Global Nuclear Energy Partnership proposal, unveiled in February 2006, envisions reprocessing as a way to reduce the amount of long-lived plutonium and highly radioactive cesium and strontium that would need to be placed in Yucca Mountain, thereby expanding its disposal capacity.[48]

The extent to which the nuclear waste issue could inhibit nuclear power expansion is difficult to assess. NRC has determined that onsite storage of spent fuel would be safe for at least 30 years after expiration of a reactor's operating license, which was estimated to be as long as 70 years. As a result, the Commission concluded that "adequate regulatory authority is available to require any measures necessary to assure safe storage of the spent fuel until a repository is available."[49] Therefore, NRC does not consider the lack of a permanent repository for spent fuel to be an obstacle to nuclear plant licensing. However, the Administration was concerned enough about repository delays to include a provision in its recent nuclear waste bill to require NRC, when considering nuclear power plant license applications, to assume that sufficient waste disposal capacity will be available in a timely manner.[50]

Six states—California, Connecticut, Kentucky, New Jersey, West Virginia, and Wisconsin—have specific laws that link approval for new nuclear power plants to adequate waste disposal capacity. Kansas forbids cost recovery for "excess" nuclear power capacity if no "technology or means for disposal of high-level nuclear waste" is available.[51] The U.S. Supreme Court has held that state authority over nuclear power plant construction is limited to economic considerations rather than safety, which is solely under NRC jurisdiction.[52] No nuclear plants have been ordered since the various state restrictions were enacted, so their ability to meet the Supreme Court's criteria has yet to be tested.

The nuclear waste issue has also historically been a focal point for public opposition to nuclear power. Proposed new reactors that have no clear path for removing waste from their sites could face intensified public scrutiny, particularly at proposed sites that do not already have operating reactors.

End Notes

1 Energy Policy Act of 1992, Title XXVIII, P.L. 102-486.
2 Seaborg, Glenn T., *The Plutonium Economy of the Future*, October 5, 1970, p. 7.
3 "World Nuclear Generation Sets Record in 2006," *Nucleonics Week*, February 15, 2007, p. 1.
4 International Energy Agency, *Monthly Electricity Survey*, January 2006.
5 Energy Information Administration, *International Energy Annual 2003*, Table 2.7; *Nucleonics Week, op. cit.*
6 *Nucleonics Week, op. cit.*; *Nuclear Engineering International*, November 2005, p. 37.
7 Nuclear Regulatory Commission, *Power Uprates for Nuclear Plants*, Fact Sheet, July 2004.
8 *Nuclear News*, "World List of Nuclear Power Plants," March 2005, p. 59.
9 Uranium Information Centre, *The Economics of Nuclear Power*, Briefing Paper 8, January2006, p. 3.
10 *Nucleonics Week*, "U.S. Utility Operating Costs, 2005," September 14, 2006, p. 7.
11 *Nuclear News*, *op. cit.*
12 See [http://www.nrc.gov/reactors/operating/licensing/renewal/applications. html]
13 *Nuclear News*, "Word List of Nuclear Power Plants," March 2006, p. 37.

[14] World Nuclear Association, *World Nuclear Power Reactors 2005-06 and Uranium Requirements*, September 21, 2006. Excludes three reactors undergoing reconstruction in the United States and Canada.

[15] 10 CFR Part 50.

[16] 54 *Federal Register* 15372, Apr. 18, 1989.

[17] After the combined license regulations were challenged in court, Congress endorsed the procedure in the Energy Policy Act of 1992 (P.L. 102-486), Title XXVIII.

[18] Weil, Jenny, "Safety of Existing Fleet to Remain the Top Priority at NRC, Klein Says," *Inside NRC*, September 4, 2006, p. 1.

[19] For more information on nuclear plant security, see CRS Report RS21131, *Nuclear Power Plants: Vulnerability to Terrorist Attack*, by Mark Holt and Anthony Andrews.

[20] See [http://www.nrc.gov/reactors/new-licensing/esp.html].

[21] *Internal Revenue Bulletin*, No. 2006-18, May 1, 2006, p. 855.

[22] 71 *Federal Register* 46306, August 11, 2006.

[23] Ibid., p. 46329.

[24] Telephone conversation with Marvin Shaw, DOE Office of General Counsel, May 16,2006.

[25] Testimony of Skip Bowman, President and Chief Executive Officer, Nuclear Energy Institute, to the Energy and Water Development Subcommittee of the House Appropriations Committee, September 13, 2006.

[26] U.S. Department of Energy, Loan Guarantee Program Office, *Federal Loan Guarantees for Projects that Employ Innovative Technologies in Support of the Advanced Energy Initiative*, Solicitation Number DE-PS01-06LG00001, August 8, 2006.

[27] EPACT refers to the cost definition in the Federal Credit Reform Act of 1990, which defines the subsidy cost as "the estimated long-term cost to the government of a direct loan or a loan guarantee, calculated on net present value basis, excluding administrative costs."

[28] U.S. Department of Energy, *Loan Guarantees for Projects that Employ Innovative Technologies; Guidelines for Proposals Submitted in Response to the First Solicitation*, effective August 8, 2006.

[29] Testimony of Skip Bowman, *op. cit.*

[30] Energy Information Administration, *Assumptions to the Annual Energy Outlook — 2006 (With Projections to 2030)*, March 2006, pp. 71-87, at [http://www.eia.doe.gov/oiaf/aeo/assumption/pdf/electricity.pdf]. The term "initiated" is not defined.

[31] The real capital charge rate is calculated based on a 6.74% discount rate. For a full discussion, see Environmental Protection Agency, *Standalone Documentation for EPA Base Case 2004 (V.2.1.9) Using the Integrated Planning Model* (September 2005), chapter 7, at [http://www.epa.gov/airmarkets/epa-ipm/bc7financial.pdf].

[32] Energy Information Administration, *Assumptions to the Annual Energy Outlook — 2006 (With Projections to 2030)*, (March 2006) p. 86, at [http://www.eia.doe.gov/oiaf/aeo/assumption/pdf/electricity.pdf].

[33] See Environmental Protection Agency, *Analyzing Electric Power Generation under the CAAA*, Office of Air and Radiation (March 1998), p. A2-12.

[34] Interdisciplinary MIT Study, *The Future of Nuclear Power*, Massachusetts Institute of Technology, 2003, p. 79.

[35] See the Nuclear Energy Institute (NEI) website at [http://www.nei.org/index.asp?catnum=1&catid=11].

[36] Patrick Moore, "Going Nuclear," *Washington Post*, April 16, 2006, p. B1.

[37] See CRS Report RS22076, *Climate Change: Summary and Analysis of the Climate Stewardship Act (S. 342, S. 1151, and H.R. 759)*, by Larry Parker and Brent Yacobucci, and CRS Report RL32755, *Air Quality: Multi-Pollutant Legislation in the 109th Congress*, by Larry Parker and John Blodgett.

[38] Ben Evans and Catherine Hunter, "Senate Rejects Global Warming Amendment," *CQ Today*, June 22, 2005.

[39] For a comparison of bills introduced in the 110th Congress, see CRS Report RL33846, *Climate Change: Greenhouse Gas Reduction Bills in the 110th Congress*, by Larry Parker.

[40] For more on S. 2724, see CRS Report RL32755, *Air Quality: Multi-Pollutant Legislation in the 109th Congress*, by Larry Parker and John Blodgett.

[41] For more on the proposal, see CRS Report RL32953, *Climate Change: Comparison and Analysis of S. 1151 and the Draft "Climate and Economy Insurance Act of 2005,"* by Brent Yacobucci and Larry Parker.

[42] For a discussion of safety valves, see CRS Report RS21067, *Global Climate Change: Controlling CO_2 Emissions — Cost-Limiting Safety Valves*, by Larry Parker.

[43] A House version of the bill, H.R. 759, has been introduced by Representatives Gilchrest and Olver.

[44] For more on the bill, see CRS Report RS22076, *Climate Change: Summary and Analysis of the Climate Stewardship Act (S. 342, S. 1151, and H.R. 759)*, by Larry Parker and Brent Yacobucci.

[45] For more on S. 150, see CRS Report RL32755, *Air Quality: Multi-Pollutant Legislation in the 109th Congress*, by Larry Parker and John Blodgett.

[46] U.S. Department of Energy, "DOE Announces Yucca Mountain License Application Schedule," news release, July 19, 2006.

[47] Data compiled by CRS. For table and details, see CRS Report RL32163, *Radioactive Waste Streams: An Overview of Waste Classification for Disposal*, byAnthony Andrews.

[48] See the Department of Energy website at [http://www.gnep.energy.gov].

[49] NRC, *Waste Confidence Decision Review*, 55 *Federal Register* 38472, Sept. 18, 1990. The 1990 decision was reaffirmed by NRC on November 30, 1999, and NRC denied a petition to amend the decision August 10, 2005.

[50] "Nuclear Fuel Management and Disposal Act," transmitted to House Speaker Nancy Pelosi and Vice President Richard Cheney March 6, 2007, by Energy Secretary Samuel Bodman.

[51] Lovell, David L., Wisconsin Legislative Council Staff, *State Statutes Limiting the Construction of Nuclear Power Plants*, October 5, 2006.

[52] Wiese, Steven M., *State Regulation of Nuclear Power*, CRS Report prepared for the House Committee on Interior and Insular Affairs, Dec. 14, 1992, p. 18.

In: Rethinking Nuclear Power in the United States ISBN: 978-1-60692-152-4
Editor: Kenneth A. Vellis

Chapter 3

COSTS OF REPROCESSING VERSUS DIRECTLY DISPOSING OF SPENT NUCLEAR FUEL*

Peter R. Orszag

Mr. Chairman, Senator Domenici, and Members of the Committee, thank you for the invitation to discuss the Congressional Budget Office's (CBO's) analysis of the costs of two alternatives for the use and disposal of nuclear fuel. For the past 50 years, the nuclear waste produced at reactors across the United States has largely been stored at the reactor sites. That practice, however, has been deemed untenable for the long run.

CBO's analysis compares the cost of two fuel-cycle alternatives for the current generation of thermal reactors. One alternative is direct disposal (as stipulated by current law), which involves using nuclear fuel once, cooling it at an interim storage site, and then disposing of it in a long-term repository. The second alternative is reprocessing, in which spent nuclear fuel is cooled and then reprocessed for one additional use in a reactor, and the wastes from reprocessing are stored in a long-term repository.

My testimony makes the following key points:

---

* This is an edited, reformatted and augmented version of a Congressional Budget Office publication dated November 2007.

- The cost of directly disposing of spent nuclear fuel is less than the cost of reprocessing it. That basic result holds across a wide range of plausible assumptions, but the magnitude of the cost difference between the alternatives varies significantly among different analyses.
- Two studies illustrate the range of estimates of the cost difference between reprocessing and direct disposal. A study by the Boston Consulting Group estimates that reprocessing spent nuclear fuel would cost $585 per kilogram—or about 6 percent more than direct disposal. CBO's analysis of another study, by a group of researchers affiliated with Harvard University's Kennedy School of Government, suggests that reprocessing would cost about $1,300 per kilogram—or more than twice as much as direct disposal.
- From its analysis of those and other studies, CBO concludes that for the roughly 2,200 metric tons of spent fuel produced each year in the United States, the reprocessing alternative would be likely to cost at least $5 billion more in present-value terms than the direct-disposal alternative over the life of a reprocessing plant. (Present-value figures convert a stream of future costs into an equivalent lump sum today.) The cost of reprocessing would be at least 25 percent greater than the cost of direct disposal.
- Major sources of uncertainty in such estimates include how much it would cost to build and operate a reprocessing facility, how long the facility would last, and the market value of reprocessed fuel.
- Policymakers evaluating the reprocessing and direct-disposal options may be concerned not only about cost but also about such potentially important issues as the impact of the alternatives on the threat of nuclear proliferation and the need for long-term storage space for spent fuel. Those issues are largely beyond the scope of CBO's analysis.

Background on Nuclear Fuel-Cycle Alternatives

As of 2006, 104 nuclear reactors were operating in the United States, with a collective generating capacity of about 100 gigawatts of electricity. Those reactors account for nearly 20 percent of the electricity produced in this country.[1]

All of the commercial nuclear power plants in the United States generate electricity by relying on the uranium-235 isotope to sustain a nuclear reaction. Uranium-235 is relatively scarce and typically makes up less than 1 percent of mined uranium ore. The bulk of that ore consists of uranium-238, which cannot be used directly to sustain a nuclear fission chain reaction. For a sustained reaction to occur, the uranium must be enriched—that is, the proportion of uranium-235 must be increased, generally to between 3 percent and 5 percent in the case of fuel for civilian reactors.

After approximately four years in a reactor, too little uranium-235 remains in the fuel to generate electricity. The spent fuel can be handled in one of two ways: Under direct disposal, it is placed in interim storage for cooling, with the goal of eventually storing it in a stable geologic formation over the long term. Under reprocessing—which is done in a few countries but not the United States—a reprocessing facility recovers the useful components of the spent fuel (uranium and certain forms of plutonium) and returns them to the fuel cycle, where they are combined with newly mined uranium to produce more reactor fuel (see Figure 1). Any waste remaining from the spent nuclear fuel after the uranium and plutonium are removed is intended to be stored in a long-term repository. Thus, under either option, some form of long-term storage facility is necessary.

No long-term repository for storing commercial nuclear waste is currently operating anywhere in the world. The Department of Energy (DOE) is planning to build and operate such a repository at Yucca Mountain in Nevada. That facility, originally scheduled to open in 1998, is now intended to start operating in 2017, although a later opening date—2020 or 2021—is more likely.[2] That date would be nearly 40 years after lawmakers directed DOE to begin studying potential sites for a deep underground repository for spent nuclear fuel.[3]

With such delays, the accumulated stock of nuclear waste is expected to exceed Yucca Mountain's mandated capacity before the facility begins accepting waste for storage. One approach to that problem is to expand the repository's capacity, either physically or by lifting the mandated limit on how much waste Yucca Mountain can accept (an option that many observers believe could be undertaken without compromising safety). Another approach is to reprocess spent nuclear fuel for reuse in reactors. That option could increase the effective capacity of the repository by allowing more nuclear waste to be stored in a given amount of space.

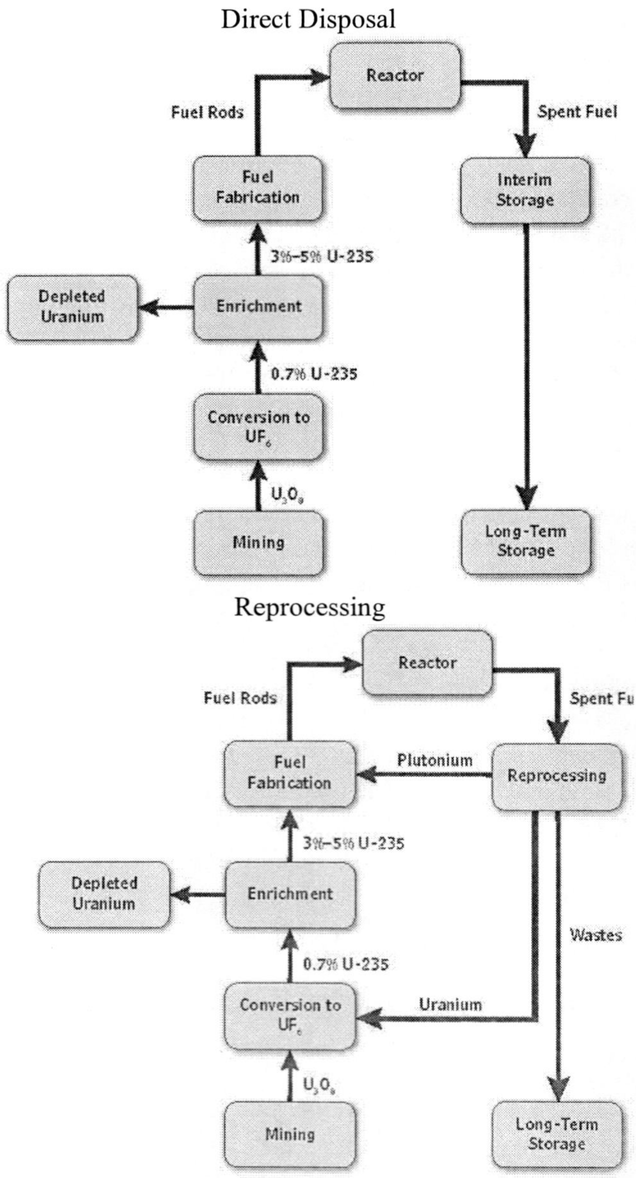

Source: Congressional Budget Office based on Ian Hore-Lacy, *Nuclear Energy in the 21st Century* (London: World Nuclear University Press, 2006).

Note: U_3O_8 = uranium oxide concentrate; UF_6 = uranium hexafloride; U-235 = uranium-235.

Figure 1. Nuclear Fuel Cycles

The main factor that determines the overall storage capacity of a long-term repository is the heat content of nuclear waste, not its volume. The waste that results from reprocessing spent fuel from thermal reactors has a lower heat content (after a period of cooling) than the spent fuel itself does. Thus, it can be stored more densely.[4] The extent of that densification directly affects the relative costs of direct disposal and reprocessing. However, unlike waste from the reprocessing process, spent fuel that has been reprocessed and used again has a higher heat content than spent fuel that has been used only once. Storing that previously recycled spent fuel in the long-term repository immediately would eliminate all of the densification benefits of reprocessing. Consequently, for reprocessing to reduce the need for— and cost requirements of—long-term storage, previously recycled spent fuel would have to be allowed to accumulate at some location outside the repository.

Besides potentially lowering long-term disposal costs, reprocessing spent nuclear fuel has the advantage of reducing expenditures for freshly mined and enriched uranium. In effect, recovering unused uranium from spent fuel extends the life of unmined uranium resources. Furthermore, recovered plutonium is not subject to many of the fuel-preparation costs (such as for conversion and enrichment) that are necessary with uranium (see Figure 1). That potential for front-end savings was especially appealing when the U.S. commercial nuclear program began in the 1950s. It became less pronounced by the 1960s, as uranium prices declined and uranium preparation techniques matured. Spot prices for uranium have recently reached historical highs (adjusted for inflation), but high prices would have to persist for decades to increase the economic viability of reprocessing.

Reprocessing and direct disposal differ not only in potential costs but also in possible risks for the proliferation of nuclear weapons. Spent nuclear fuel itself poses little risk of proliferation because the plutonium it contains is mixed with highly radioactive elements and can be recovered only in dedicated reprocessing facilities.[5] But the most widely used method of reprocessing—called plutonium and uranium recovery by extraction, or PUREX—yields pure plutonium, which has relatively low radioactivity and can be handled directly. Thus, the PUREX method recovers plutonium in a form that poses risks for theft and proliferation. Other reprocessing methods being considered by policymakers try to reduce those risks by not separating pure plutonium from spent fuel.

Reprocessing Facilities

The United States has limited experience with commercial reprocessing. Three reprocessing plants were built for commercial use, but only one—a plant in West Valley, New York, that opened in 1966—achieved any level of operation. The need for costly upgrades caused it to close in 1976, having handled only spent fuel from national defense operations.[6]

Today, five nations—France, the United Kingdom, Japan, Russia, and India—have or are developing reprocessing facilities. The world's largest reprocessing plant is located in La Hague, France, and has a gross capacity of 1,700 metric tons per year. The United Kingdom has two reprocessing centers at its Sellafield Nuclear Site: a 900-metric-ton thermal oxide reprocessing plant (THORP) and a facility that specializes in reprocessing waste for two specific British nuclear facilities (Oldbury and Wylfa, both of which are expected to cease operations by 2010). Another reprocessing facility has been under construction in Rokkasho, Japan, since the late 1980s. The start of commercial operations there has been delayed several times but is now expected to occur later this year.

Thermal Reactors Versus Fast-Neutron Reactors

CBO's analysis compares the cost of reprocessing nuclear fuel from thermal reactors—the type of commercial reactor used now in the United States—with the cost of using uranium fuel a single time and then putting all of it in a geologic repository. However, some policy initiatives, such as the Global Nuclear Energy Partnership, have focused on another type of reactor: an advanced burner reactor, which is a type of fast-neutron reactor. Whereas thermal reactors rely on less energetic, or modulated, neutrons to sustain a nuclear chain reaction, fast-neutron reactors rely on unmodulated (and hence more energetic) neutrons for a reaction. Fast-neutron reactors use plutonium as a fuel source rather than uranium because plutonium maintains a reaction with unmodulated neutrons more readily than commercial-grade enriched uranium does.

Fast-neutron reactors offer several advantages. They can convert plentiful uranium-238 (which is not usable for nuclear chain reactions) into plutonium in such a way as to produce (or breed) more plutonium than the reactor itself uses. In that way, a fast-neutron breeder reactor can extend uranium resources by accessing 60 to 100 times more of the energy content of uranium than thermal reactors can.[7] Fast-neutron reactors also generate less spent fuel than thermal reactors do. Besides uranium and plutonium, spent nuclear fuel includes two other types of waste: fission products and minor actinides. Minor actinides decay less rapidly than fission products do. Because the capacity of a geologic repository depends to a significant degree on the long-term radioactivity of waste, it is greatly influenced by the amount of minor actinides present in spent fuel. Advanced burner reactors can potentially burn all of the actinides in nuclear fuel, so waste from those reactors requires less geologic storage space than does either spent nuclear fuel from thermal reactors or the waste from reprocessing thermal reactors' spent fuel.

Whereas reprocessing spent fuel is merely an option with thermal reactors (to extend uranium resources or to potentially expand long-term storage capacity), it is an integral part of the fuel cycle for advanced burner reactors. The fuel needed to power advanced burner reactors can be collected by reprocessing spent fuel from thermal reactors or from burner reactors. Furthermore, if burner reactors are used to reduce thermal-reactor waste, spent nuclear fuel must be reprocessed.

This testimony does not consider reprocessing in the context of fast-neutron reactors, for three reasons. First, no commercial fast-neutron reactors exist in the United States and none are planned. Second, the 60-year-old PUREX process is essentially the only reprocessing method now used for thermal reactors, and given its long history, the cost of PUREX is better known than the costs of more-recent reprocessing technologies that are being considered for fast-neutron reactors. Third, reprocessing fuel for advanced burner reactors would probably require reprocessing nuclear waste from thermal reactors as a first step to create the fuel for the burner reactors and to manage any existing thermal-reactor waste. Thus, reprocessing thermal-reactor waste can be thought of as a transitional element to a burner-reactor program.

Cost Comparisons for Direct Disposal and Reprocessing

Reprocessing nuclear fuel could have several economic advantages over direct disposal. It could reduce spending on newly produced uranium fuel and extend the useful life of uranium resources. In addition, it could save money on long-term storage by reducing the size of the repository necessary to handle spent nuclear fuel or by delaying the need to expand such a facility in the future.

With current reactor technology, reprocessing would also have economic disadvantages. First, it would require building dedicated facilities to recover the useful components of spent nuclear fuel and then to combine them into a form usable in a nuclear reactor. Second, previously recycled spent fuel would also need some form of long-term storage.

To quantify the relative costs of reprocessing and direct disposal, CBO's analysis focuses on the costs of handling nuclear fuel after it is discharged from a reactor. In the case of reprocessing, those costs include the costs of reprocessing services (both recovering uranium and plutonium and fabricating them into usable nuclear fuel), transportation, and long-term disposal of wastes, partially offset by "fuel credits," which various models use to reflect the value of the reprocessed fuel (in the form of savings on the costs of newly purchased fuel). In the case of direct disposal, the costs in this analysis include costs for interim storage to cool the spent fuel, transportation, and long-term disposal.

CBO reviewed a number of studies that shed light on the costs of nuclear fuel-cycle alternatives, including reports by the National Research Council, the Massachusetts Institute of Technology, the Idaho National Laboratory, and the Nuclear Energy Agency of the Organisation for Economic Co-operation and Development (OECD).[8] However, CBO's analysis focused on two studies in particular: a 2006 report by the Boston Consulting Group (BCG) and a 2003 report by researchers at Harvard University's Kennedy School of Government.[9] Those two studies are the only recent analyses available that investigate the costs of all facets of both reprocessing and direct disposal (including transportation, interim storage, and credits for recycled fuel). Other studies consider only the costs of reprocessing or do not examine the various components of total costs. In addition, the two studies' estimates of the cost of reprocessing services—one of the largest cost elements— bound the range of estimates provided in, or implied by, the other studies. The Kennedy study's estimate of the cost of reprocessing services is about twice the size of the BCG study's estimate. Other studies that CBO examined had cost estimates for reprocessing services that fell within the range defined by those two reports.

Evaluating the Boston Consulting Group and Kennedy Studies on a Common Ground

The BCG study concludes that reprocessing spent fuel costs about $30 more per kilogram than direct disposal (which the study estimates at $555 per kilogram). To directly compare that estimate with the results of the Kennedy study, CBO modified the Kennedy study to reflect a similar initial framework as in the BCG study. In that framework, the Kennedy study implies that reprocessing costs about $700 more per kilogram than direct disposal. Given the volume of waste expected to be generated over the lives of the plants evaluated, those estimates suggest that the present-value cost of reprocessing exceeds that of direct disposal by about $2 billion for the BCG study and by about $26 billion for the Kennedy study, as modified by CBO. (Present-value calculations use a discounted cost framework that describes the amount of funds that would be necessary in 2007 to pay all of the costs of a waste-management option over the assumed lifetime of a reprocessing plant.)

Several differing assumptions account for much of the gap between those two present-value estimates. Such assumptions include the interest rate used to estimate the present value of future costs (the discount rate), the relationship between a reprocessing plant's yearly operating costs and total capacity costs, the time horizon over which the plant operates, the cost of a long-term repository, and the degree to which waste from reprocessing can be stored more densely than spent nuclear fuel in the repository. Changes to any of those assumptions will affect the relative costs of the two waste-handling alternatives. To control for those differences, CBO's analysis imposed a common set of cost assumptions on the estimates from the BCG study and from the modified Kennedy study. In particular, CBO assumed the following:

- A discount rate of 3.5 percent, which lies between the rates used in the two studies.
- Plant operating costs that equal 6 percent of the plant's capital costs, a rule of thumb adopted in an analysis by OECD's Nuclear Energy Agency.[10] That figure lies between the 4.6 percent ratio implied by the BCG study and the 7.5 percent ratio implied by the Kennedy study.
- A lifetime of 40 years for a reprocessing plant, the midpoint between the 50-year figure used in the BCG study and the 30-year lifetime assumed in the Kennedy study.

- Repository costs of $1,036 per kilogram of heavy metal stored in the repository, an estimate that CBO developed using cost data from DOE. That cost exceeds both the $736 per kilogram figure in the BCG study and the $868 per kilogram estimate in the Kennedy study.
- A densification factor of 2.5 applied to repository capacity, based on the study by the Idaho National Laboratory.[11] That figure is between the factor of 4 used in the BCG study and the factor of 2 implied by the Kennedy study.

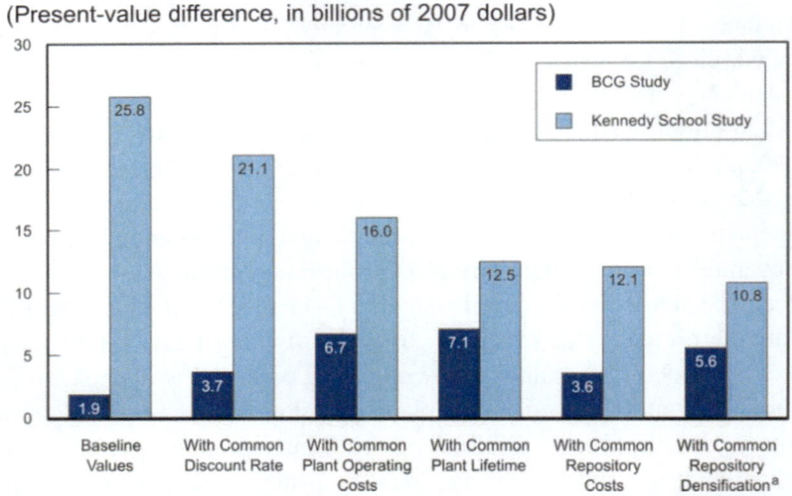

Source:Congressional Budget Office based on Boston Consulting Group (BCG), *Economic Assessment of Used Nuclear Fuel Management in the United States* (study prepared for AREVA Inc., July 2006); and Matthew Bunn and others, *The Economics of Reprocessing vs. Direct Disposal of Spent Nuclear Fuel* (Cambridge, Mass.: Belfer Center for Science and International Affairs, John F. Kennedy School of Government, Harvard University, December 2003).

Note:The numbers shown here represent the extent to which the future costs of reprocessing spent nuclear fuel exceed the future costs of direct disposal on a present-value basis. (Present-value figures convert a stream of future costs into an equivalent lump sum today.) The first set of bars shows the values implied by the BCG and Kennedy School studies under their different assumptions. The successive sets of bars show the cumulative effects of applying common assumptions to the two baseline estimates. The common assumptions that CBO used for this analysis are described in detail in the text.

a. Densification is the extent to which waste from reprocessing can be stored more densely than unreprocessed spent fuel in a long-term repository.

Figure 2. Discounted Cost Differences Between Reprocessing and Direct Disposal

As those common assumptions are applied successively, the two present-value estimates of the difference between reprocessing and direct-disposal costs narrow from a range of $2 billion to $26 billion to a range of $5 billion to $11 billion (see Figure 2).

Most of that remaining gap is attributable to the two studies' different assumptions about the costs of building and operating a reprocessing plant. The BCG study estimates construction costs at about $17 billion for a plant with a capacity of 2,500 metric tons per year. A meaningful comparable estimate cannot be derived from the Kennedy study because that analysis does not explicitly differentiate between capital and operating costs. The likelihood that a newly built U.S. plant would match either study's cost assumptions is difficult to judge; the historical record provides scant evidence about the overall cost of a reprocessing facility and its component parts. Not only are there few large-scale commercial reprocessing plants, but only limited information is available about their construction and operating costs.

Neither the 900-metric-ton THORP facility in the United Kingdom nor the 1,700-metric-ton La Hague facility in France has enough capacity to handle the 2,200 metric tons of nuclear waste generated in the United States each year—the amount considered in this analysis. Thus, a facility larger than any past or current example would be necessary if a single reprocessing plant was to handle the United States' entire annual output of spent nuclear fuel.

A larger facility would be more costly than existing plants, although to what degree is unknown. The limited information available suggests that the THORP plant cost around $6.3 billion to build (in 2007 dollars). The BCG study indicates that the construction cost of the La Hague facility was around $18 billion (unlike the THORP estimate, however, that total includes a fabrication facility for recycled fuel, which increases the overall cost). The nearly complete 800-metric-ton Rokkasho facility will reportedly cost about $21 billion, but part of that cost is attributable to specifics of the plant's location that would not necessarily apply to a U.S. facility. Given the lack of numerous commercial reprocessing facilities to use as examples, it is difficult to know how much geographic location, economies of scale, and regulatory environment affect the cost of a reprocessing plant.

All of the costs for reprocessing services included in this analysis assume that the plant would operate near capacity for its entire life. History, however, suggests that such an assumption might be optimistic and therefore that the unit cost of reprocessing could be higher than described here. Neither THORP nor La Hague has operated close to full capacity for a substantial period. THORP has been closed for more than two years after experiencing a radioactive leak. Before that, the plant operated at about 60 percent of capacity over its first 11 years. Although La Hague has not had the technical problems of the THORP facility, it too is operating well below full capacity: at approximately 65 percent, according to recent estimates. Operating at less than full capacity limits the amount of spent fuel that can be handled for a given cost.

Another factor that could increase the estimated cost of reprocessing relative to direct disposal is the discount rate used in present-value calculations. The rate assumed in this analysis is similar to those used in the BCG and Kennedy studies and slightly above a risk-free government rate—but it is well below the rate that might be applied for this type of project. A higher discount rate would result in a larger cost difference between reprocessing and direct disposal.

The relative cost of reprocessing is also affected by the market value of recycled fuel. As noted above, the fuel credits used in this analysis reflect front-end savings from using recycled fuel rather than newly mined uranium. If the costs of uranium mining and fuel preparation increased, and if recycled fuel proved to be a good substitute for newly mined uranium in nuclear reactors, higher fuel credits could offset the cost of reprocessing to a greater extent. Although uranium prices are currently high by historical standards, it is not certain whether high prices will continue in the future or whether current prices will encourage additional uranium development that could lower prices. Furthermore, modifying a nuclear reactor to use recycled fuel entails some costs, which would offset a portion of the potential fuel credits from reprocessing.

Sensitivity to Varying Assumptions

Although the size of the cost difference between reprocessing and direct disposal depends on inputs to specific models, the conclusion that reprocessing is more expensive than direct disposal generally applies under various assumptions. CBO tested the sensitivity of the results to changes in some of the key parameters of this analysis.

- An increase of 1 percentage point in the discount rate increases the difference in present-value costs between reprocessing and direct disposal by between $3 billion and $4 billion.

- A reduction in the assumed operating costs of a reprocessing plant narrows the cost gap between reprocessing and direct disposal. For example, decreasing the ratio of a plant's operating costs to its capital costs by 1 percentage point reduces the present-value cost differential by between $2 billion and $3 billion. However, operating costs would have to be at least 50 percent lower for reprocessing to cost the same as or less than direct disposal.
- A change in the assumed operating lifetime of a reprocessing facility has no material impact on the cost differential for the two waste-handling alternatives.
- A rise in the cost of the long-term storage repository reduces the difference between the costs of reprocessing and direct disposal. That cost would have to increase to a very great extent, however, for direct disposal to cost as much as reprocessing. Even then, if the factors responsible for the increase (such as general growth in materials and construction costs) also applied to the cost of a reprocessing plant, reprocessing would continue to have a cost disadvantage.
- An increase in the extent to which waste from reprocessing can be stored more densely than unreprocessed spent fuel (the densification factor) lowers the cost of reprocessing relative to direct disposal. However, reprocessing remains at a cost disadvantage under plausible values for densification.

In conclusion, the cost of reprocessing may be comparable to that of direct disposal under limited circumstances, but under a wide variety of assumptions, reprocessing is more expensive (given current reactor technology).

Policymakers weighing the merits of reprocessing and direct disposal may have other concerns besides cost—such as extending U.S. uranium resources, reducing the threat of nuclear proliferation by adopting advanced burner technologies, or lessening the demand for long-term storage space. Judging whether those goals justify the added costs of reprocessing is ultimately a decision for policymakers.

End Notes

[1] Department of Energy, Energy Information Administration, *Annual Energy Review 2006* (June 2007), Table 8.2a, available at www.eia.doe.gov/emeu/aer/pdf/pages/sec8_8.pdf.

[2] Statement of Edward F. Sproat, Director, Office of Civilian Radioactive Waste Management, at the 178th meeting of the Nuclear Regulatory Commission's Advisory Committee on Nuclear Waste, April 10, 2007, available at www.nrc.gov/readingrm/doc-collections/acnw/tr/2007/ nw041007.pdf

[3] As originally enacted, the Nuclear Waste Policy Act of 1982 called for studies of three potential sites for long-term geologic repositories. Sections 5011 and 5012 of the Nuclear Waste Policy Amendments Act of 1987 effectively cancelled any investigation into sites other than the one at Yucca Mountain.

[4] The "densification factor" describes that relationship; for example, a densification factor of 2 indicates that twice as much waste from reprocessing can be stored at the same total cost (in other words, that the unit cost of storage is half as much).

[5] Steve Fetter and Frank N. von Hippel, "Is U.S. Reprocessing Worth the Risk?" *Arms Control Today* (September 2005), available at www.armscontrol.org/act/2005_09/Fetter-VonHippel.asp.

[6] Anthony Andrews, *Nuclear Fuel Reprocessing: U.S. Policy Development,* Report for Congress RS22542 (Congressional Research Service, November 29, 2006).

[7] Uranium Information Centre, *Fast Neutron Reactors,* Briefing Paper No. 98 (Melbourne, Australia: Australia Uranium Association, June 2006), available at www.uic.com.au/nip98.htm.

[8] National Research Council, *Nuclear Wastes: Technologies for Separations and Transmutations* (Washington, D.C: National Academy Press, 1996); Massachusetts Institute of Technology, *The Future of Nuclear Power* (Cambridge, Mass.: MIT, 2003), available at http://web.mit.edu/ nuclearpower; D.E. Shropshire, *Advanced Fuel Cycle Cost Basis* (Idaho Falls: Idaho National Laboratory, April 2007), available at www.inl.gov/technicalpublications/Documents/3667084. pdf; and Nuclear Energy Agency, *The Economics of the Nuclear Fuel Cycle* (Paris: Organization for Economic Co-operation and Development, 1994), available at www.nea.fr/html/ndd/ reports/efc.

[9] Boston Consulting Group, *Economic Assessment of Used Nuclear Fuel Management in the United States* (study prepared for AREVA Inc., July 2006); and Matthew Bunn and others, *The Economics of Reprocessing vs. Direct Disposal of Spent Nuclear Fuel* (Cambridge, Mass.: Belfer Center for Science and International Affairs, John F. Kennedy School of Government, Harvard University, December 2003).

[10] Nuclear Energy Agency, *Accelerator-Driven Systems and Fast Reactors in Advanced Nuclear Fuel Cycles: A Comparative Study* (Paris: Organization for Economic Co-operation and Development, 2002), p. 211, available at *www.nea.fr/html/ndd/reports/2002/* nea3109.html.

[11] Shropshire, *Advanced Fuel Cycle Cost Basis*,p.L-12.

In: Rethinking Nuclear Power in the United States ISBN: 978-1-60692-152-4
Editor: Kenneth A. Vellis

Chapter 4

THE FEDERAL GOVERNMENT'S LIABILITIES UNDER THE NUCLEAR WASTE POLICY ACT*

Kim Cawley

Mr. Chairman, Congressman Ryan, and Members of the Committee, I am pleased to appear before you today to discuss issues related to the federal government's liability under the Nuclear Waste Policy Act of 1982 (NWPA).

My testimony today makes the following points:

- By law, the federal government is responsible for permanently disposing of spent nuclear fuel generated by civilian facilities, which pay fees for that waste disposal service. Regardless of how the government meets that responsibility, discharging those liabilities will require significant federal spending over many decades.
- The Nuclear Waste Policy Act authorized a system to manage radioactive waste, including an underground repository to permanently dispose of spent nuclear fuel from civilian facilities. Currently, the federal government is 10 years behind schedule in its contractual obligations to remove and dispose of such waste; by the time the repository might be opened, it is likely to face at least a 20-year waste backlog.

---

* This is an edited, reformatted and augmented version of a Congressional Budget Office publication dated October 2007.

- In the absence of a federal underground repository to accept nuclear waste for storage, taxpayers are now starting to pay—in the form of legal settlements with utilities—for a decentralized waste storage system at sites around the country. (Those payments are being made from the Department of the Treasury's Judgment Fund.) The Department of Energy (DOE) currently estimates that payments to utilities pursuant to such settlements will total at least $7 billion, and possibly much more if the program's schedule continues to slip. Regardless of whether or when the government opens the planned repository, those payments are likely to continue for several decades.
- Ultimately, the repository that is now authorized under NWPA will not provide sufficient capacity to store all of the waste for which the federal government is responsible. The statutory cap on the amount of waste that can be stored there is significantly lower than the volume of waste that DOE expects will be generated during the lifetimes of existing nuclear facilities, let alone the additional volume from any new facilities that may be built. Without a change in law to allow construction of disposal facilities with sufficient capacity to accommodate all of the waste that will be generated, taxpayers will need to pay utilities to dispose of a substantial amount of additional waste in the future.
- Contractual liabilities associated with nuclear waste from civilian power plants are one component of the government's broader liabilities for remedying environmental contamination, much of which results from operating the nation's nuclear weapons complex.

The Federal Government's Responsibilities under the Nuclear Waste Policy Act

The Nuclear Waste Policy Act gave the federal government statutory responsibility for permanently disposing of spent nuclear fuel generated at civilian nuclear reactors and for disposing of radioactive waste generated as a result of federal activities related to the manufacturing of nuclear weapons. Under the law, the federal government, through the Department of Energy, faces substantial costs to establish a federal repository for the nation's nuclear waste. It has also incurred contractual obligations to remove waste from civilian nuclear facilities.

Under NWPA, the federal government will have to spend tens of billions of dollars over many decades to fulfill its obligations to dispose of waste from the current generation of civilian nuclear reactors. The government will also be responsible for waste from any new facilities that may be brought online in the future. However, the waste from those new facilities cannot be accommodated in the repository envisioned under NWPA because of statutory constraints on the amount of waste the repository can store.

The Federal Repository Authorized for Yucca Mountain

The Nuclear Waste Policy Act authorized DOE to build a geologic repository to permanently store up to 70,000 metric tons of spent nuclear fuel generated by civilian nuclear power plants and high-level radioactive waste generated by federal facilities. Under current law, Yucca Mountain in Nevada is the only place where such a repository may be located. To proceed with construction and operation of the facility, DOE must apply for and receive a license from the Nuclear Regulatory Commission (NRC). DOE expects to submit that application for authority to begin constructing the repository in 2008. (To date, many of DOE's activities have focused on analyses required to support the license application.) If the NRC approves the application within three years and if other key regulatory requirements are subsequently met, DOE expects the planned repository at Yucca Mountain to begin accepting waste in 2017, although the department has recently indicated that this planned schedule could slip further.[1]

The law also addressed how the disposal of spent nuclear fuel and defense-related waste was to be paid for. Under NWPA, the costs for disposing of the waste are to be borne by the parties that generate it, and the law authorizes DOE to levy fees on the nuclear power industry to cover those costs. The law also authorizes appropriations from the Treasury's general fund to pay for disposing of high-level radioactive waste generated by the nation's defense programs.

The cash flows since 1983 for major components of the nuclear waste disposal program (summarized in Table 1) are described in the following sections.

Table 1. Federal Cash Flows to and from the Nuclear Waste Fund

(Billions of nominal dollars)		
	Cumulative Totals, 1983 Through 2006	**CBO Estimate, 2007**
Deposits		
Annual fees	13.3	0.8
One-time fees	1.5	0
Subtotal, fees	14.8	0.8
Interest credited[a]	10.9	0.8
Total, Deposits	25.7	1.6
Disbursements	6.7	0.2
Balance	19.0	20.4[b]
Memorandum:		
Appropriations for Defense-Related Activities Budget authority	3.1	0.3

Sources: Congressional Budget Office; Department of Energy.

a. Intragovernmental transfers from general revenues.

b. Cumulative fund balance.

Financing the Cost of Disposing of Civilian Nuclear Waste

Starting in 1983, NWPA authorized DOE to charge electric utilities fees to cover the cost of disposing of the nuclear waste they generate. Utilities today pay annual fees at a rate of 1 mil (0.1 cent) per kilowatt-hour of electricity generated and sold by nuclear power plants. The fees, which are recorded in the budget as mandatory offsetting receipts (a credit against direct spending), are deposited into the Nuclear Waste Fund, or NWF (a special fund in the Treasury that records cash flows associated with the civilian nuclear waste program). Amounts in that fund are available for spending only to the extent provided in annual appropriation acts. Under NWPA, DOE is required to periodically review and, if necessary, adjust those fees to ensure that the fund has sufficient resources to pay for disposing of the utility industry's nuclear waste. DOE has not increased that annual charge since 1983.

In addition to the ongoing yearly fees, NWPA established one-time fees to cover the cost of disposing of waste that was generated before the law was enacted. DOE provided utilities with several options for paying that one-time charge, but several utilities have not yet paid the fee, and a significant amount remains uncollected.

The Nuclear Waste Policy Act authorized appropriations from the Nuclear Waste Fund to cover the costs of the civilian nuclear waste program and also permitted DOE to borrow from the Treasury (subject to approval in appropriation acts) if balances in the fund were insufficient to cover the program's immediate costs. (The law stipulates that amounts borrowed from the Treasury must be repaid from future fee collections.) In addition, the law authorized the Secretary of the Treasury to invest the fund's unspent balances in nonmarketable Treasury securities, which are credited with interest.

As shown in Table 1, $25.7 billion has been credited to the NWF from its inception in 1983 through the end of fiscal year 2006. That amount includes fees paid by the nuclear industry totaling $14.8 billion as well as $10.9 billion from intragovernmental transfers of interest earnings. Cumulative expenditures from the fund during that period totaled about $6.7 billion, mostly for analyses related to the waste disposal program and for appropriations to DOE for initial design work on the Yucca Mountain facility. The NRC and other federal entities also received modest appropriations from the fund for work related to the program, leaving an unspent balance of about $19.0 billion at the end of 2006. The Congressional Budget Office (CBO) estimates that in 2007, another $1.6 billion was credited to the fund—half of which came from fees and half from interest. Expenditures in 2007 totaled $0.2 billion, bringing the fund's current balance to $20.4 billion, in CBO's estimation.

If all of today's 104 licensed nuclear reactors continue to generate electricity, future annual receipts from industry fees are likely to average between $750 million and $800 million. Most U.S. nuclear power plants began operating in the mid-1970s or during the 1980s under 40-year licenses. The NRC has approved 20-year extensions to the licenses of nearly half of the plants in operation today, and it anticipates that many of the others will apply for such licenses. When those plants reach the end of their license extensions (or their economically useful lives) and cease operations—probably in the 2030s and 2040s—they will pay no additional fees to the Nuclear Waste Fund to cover long-term costs related to their waste.

Receipts from the one-time fees that remain unpaid and become due once the federal repository is opened currently amount to about $3 billion, DOE estimates.[2] Interest accrues on the balances due from those one-time fees until the utilities pay them to the government. Also accruing and adding significantly to the fund's balances are credits of interest on the fund's unspent dollars. Those amounts are intragovernmental transfers and do not create net receipts to the federal government. However, they do add to the resources that are authorized to be used for the waste disposal program.

Financing the Costs Associated with Defense-Related Nuclear Waste

In addition to the amounts appropriated from the fees and interest credited to the NWF, the Congress has made annual appropriations to the nuclear waste program to cover the costs that DOE estimates are related to the disposal of nuclear waste generated by federal defense programs. In 2001, DOE determined that just over one-fourth of the total costs of the waste disposal program are attributable to the disposal of defense-related nuclear waste and that this share of the program's total costs should be paid for with appropriations from the general fund of the Treasury.[3] Since 1993, the Congress has provided nearly $3.4 billion from the general fund for such costs.

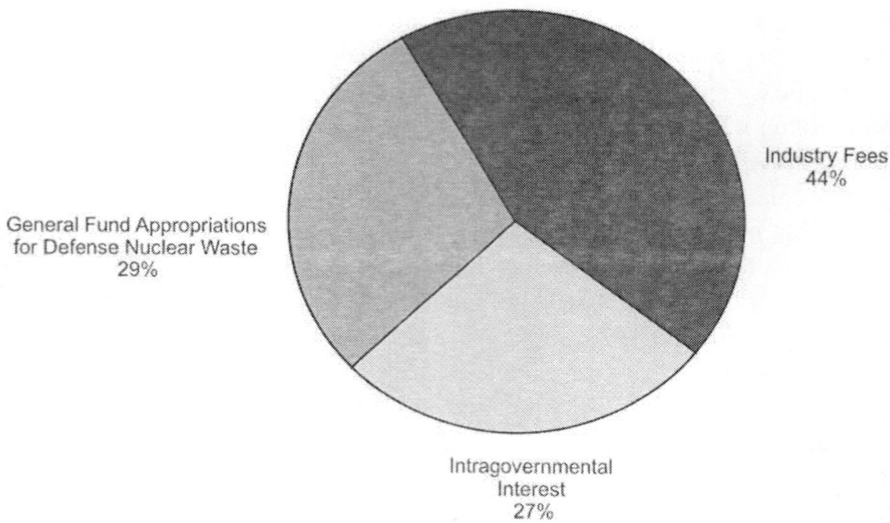

Source: Department of Energy.
Note: Costs were calculated as 2000 dollars.

Figure 1. Projected Sources of Funds for Total Life-Cycle Costs of the Nuclear Waste Disposal Program

Estimates of Total Life-Cycle Costs

In 2001, DOE published an estimate of the total costs—including transportation and project management—associated with the planned underground nuclear waste disposal facility. In DOE's estimation, the project would cost $57.5 billion in 2000 dollars (with an associated range of accuracy of plus or minus 40 percent) over an operating period of more than 100 years.[4] DOE estimated that nearly three-fourths of the facility's total life-cycle costs would be attributable to waste generated by civilian facilities; the remaining portion would be attributable to defense-related waste.

DOE also published a study in 2001 reporting on whether the annual fee charged for nuclear waste disposal would generate enough money in the Nuclear Waste Fund—including the interest anticipated on unspent fees—to pay for the estimated life-cycle costs associated with disposing of civilian waste.[5] In that study, key findings of which are summarized in Figure 1, DOE estimated that fees paid by commercial nuclear power plants would cover 44 percent of the program's total costs, and interest credited to unspent NWF balances of fees would cover 27 percent. DOE assumed that appropriations for costs attributable to defense-related waste would cover the remaining 29 percent of total costs.

Using its 2001 program design as a reference case, DOE determined that the annual fee (plus accrued interest) was likely to generate sufficient balances to cover the estimated costs of civilian waste disposal. The agency reaffirmed that determination in 2002 but has not completed an updated study of fee adequacy since that time. [6]

Judgments about whether the fee is adequate are highly sensitive to estimates of certain key variables, such as project costs and inflation. Determinations of adequacy are also sensitive to estimates of the interest credited to the fund—estimates that are a function of interest rates and fund balances, which in turn depend on projections of appropriated spending from the fund. In CBO's view, DOE's 2001 and 2002 analyses used reasonable economic assumptions, but it is unclear whether other assumptions— particularly those related to the program's scope and costs— are still an appropriate basis for determining the adequacy of utilities' annual fees. For example, because many plants have received 20-year license extensions, the total volume of waste that those plants are expected to generate—and thus the scope of a federal program necessary to manage it—have increased significantly since 2001. An updated analysis would need to take that factor into account, as well as the impact of escalating construction costs and delays in the scheduled opening of a federal repository. The Government Accountability Office, looking back, found that many DOE construction projects involving complex technologies had exceeded their original estimates of costs and experienced delays in their schedules.[7]

Federal Contractual Obligations and Liabilities for Nuclear Waste

Under contracts signed with electric utilities pursuant to NWPA, DOE was scheduled to start removing waste from storage sites at individual power plants for transport to a federal storage or disposal facility by 1998. DOE documents suggest that the planned underground nuclear waste disposal site at Yucca Mountain will not be ready to accept waste before 2017—or nearly 20 years late.[8] Moreover, that schedule assumes that legislation, which has not been enacted, will make changes to the nuclear waste program—in particular, to simplify key regulatory procedures. Without those changes, according to DOE, the actual date that the agency begins to accept waste is likely to be even later.

After the federal government missed its 1998 contractual deadline to start collecting waste, electric utilities began—successfully—to sue the government for damages incurred as a result of the agency's failure to meet that deadline. In seeking to resolve the initial lawsuits, DOE anticipated that it would pay court-awarded damages to individual utilities from amounts appropriated from the Nuclear Waste Fund or by issuing credits to those utilities (to reduce their future fee payments to the fund) in the amount of the damages that had been awarded.

In 2002, however, the U.S. Court of Appeals for the Eleventh Circuit held that DOE could not use the Nuclear Waste Fund to pay the damages resulting from the government's breach of its contracts.[9] According to the court, the costs of interim storage incurred by the utilities because of the breach were not within the uses of the fund that were permitted under NWPA. Also, the court pointed out, the department would inevitably raise future fees to compensate for any such payments—so that the injured utilities would ultimately bear the costs of the contract breach if they were paid from the fund. In addition, utilities that did not litigate their claims would end up paying larger fees to cover the cost of damage claims made by other utilities. Agreeing with the parties that brought the lawsuit, the court stated that making utilities contribute to a fund that disproportionately paid the storage costs of other utilities would raise a serious constitutional "takings" question. Following the court's decision, the government subsequently paid damages to the utilities from the Treasury's Judgment Fund.

The Judgment Fund

The Judgment Fund is a permanent, indefinite appropriation from the Treasury that is available to pay final judgments and awards against the United States that cannot legally be paid from any other existing appropriation.[10] (The fund has no fiscal year limitations, and there is no need for the Congress to appropriate money to replenish it.) The fund provides the authority for the government to pay for most court judgments and settlement agreements entered into by the Department of Justice to resolve actual or imminent lawsuits against the federal government. Generally, agencies are not required to reimburse the Judgment Fund for payments made on their behalf unless the Congress appropriates money specifically for that purpose.

Judgments Awarded and Paid to Utilities under the Nuclear Waste Policy Act

Under the Department of Justice's settlements with electric utilities, utilities have been reimbursed for the actual costs they incurred because of DOE's partial breach of its contracts. Such costs are unique to each nuclear power plant and depend partly on the age and operating status of the plant and the size and configuration of the plant's available space for nuclear waste storage.

The Judgment Fund has paid $290 million to four electric utilities as compensation for the costs they incurred because the federal government could not begin to accept nuclear waste for disposal in 1998. That amount includes a payment of $35 million to the federally owned Tennessee Valley Authority. The government has appealed (or may appeal) judgments worth another $337 million. Five additional trials for damages have been completed and are awaiting judgments, and 44 other cases have not yet been tried. Because judicial claims for damages are made retrospectively, many more cases can be expected in the coming decades as utilities seek to recover their ongoing costs for storing nuclear waste long after they expected it to be removed to a permanent disposal site.

Future Settlements under the Nuclear Waste Policy Act

Litigation is ongoing regarding how to calculate damages for DOE's partial breach of its contractual commitments. DOE currently estimates that if the agency begins to accept waste in 2017, taxpayers' total liabilities to electric utilities will total roughly $7 billion (in today's dollars). Further, it anticipates that payments from the Judgment Fund will span a number of decades after 2017.[11]

DOE's estimate of future damages is uncertain and is predicated on the department's views of the types of additional business and waste storage expenses that the courts will determine are appropriate and reasonable and should be paid by DOE. Those determinations will depend on such factors as the estimated rate at which DOE would have removed waste from a particular facility if the agency had been able to accept waste in 1998. If utilities successfully argue that the waste-acceptance rate used for the purpose of calculating damages should exceed the rate used in DOE's projections of liabilities, costs would probably surpass $7 billion.

Similarly, costs may be greater if the courts take a broader view of the types of expenses for which utilities should be compensated. Although the federal government is responsible for the permanent disposal of nuclear waste, individual utilities are responsible for storing the waste until it can be delivered to a permanent storage facility. Because the site characteristics of individual utilities vary, the determination of incremental expenses incurred at particular sites must be made on a case-by-case basis and will ultimately depend on the courts' views, which could differ from DOE's.

Estimates of liabilities will increase if the schedule for completing the planned repository slips further and waste continues to accumulate at utilities' storage sites. For example, according to DOE, estimated liabilities will increase from $7 billion to $11 billion if the agency starts accepting waste in 2020.[12] And even then, it will face a backlog that, at best, will take more than 20 years to eliminate. As long as the agency remains behind schedule, taxpayers will continue to incur liabilities.

The Outlook for the Federal Government's Liabilities

The Nuclear Waste Policy Act sets the storage capacity of the Yucca Mountain site at no more than 70,000 metric tons. DOE estimates that roughly 65,000 metric tons of existing spent nuclear fuel and high-level defense waste are currently slated for disposal there. The nation's existing nuclear power plants generate another 2,000 metric tons of waste per year. If they continue to produce waste at that rate, the total volume will exceed Yucca Mountain's statutory capacity within a few years, well before the repository is scheduled to open.

As a result, without a change in law to expand the Yucca Mountain facility or designate another site for disposal, there will be insufficient capacity to dispose of wastes generated over the lifetimes of the nuclear power plants that are operating today. The resulting waste storage compensation payments to utilities from the Judgment Fund for waste that cannot be permanently disposed of would add significantly to federal liabilities.

Moreover, the NRC has announced that it expects to receive applications for licenses to build 32 new nuclear power plants in the next few years. If constructed, each of those plants would produce around 20 metric tons of waste per year, or about 1,000 metric tons over a 40-to 60-year period of operations. Such plants would also pay fees to the Nuclear Waste Fund, and their waste would become a federal liability because under NWPA, nuclear plants are required to sign waste disposal agreements with DOE. Without additional storage capacity, that waste could become an additional liability of the Judgment Fund.

Accounting and Budgeting for Federal Liabilities

The federal government issues two different reports on its fiscal performance: the budget and *The Financial Report of the United States Government*. The budget largely measures cash flows in and out of the Treasury. The financial report, by contrast, primarily uses an accrual basis of accounting to measure assets, liabilities, revenues, and expenses. The principal difference between accrual and cash accounting is the timing of the recognition of transactions: An accrual system generally recognizes them when an economic event occurs rather than when the resulting cash flows take place.[13]

On the federal balance sheet, liabilities reflect obligations of the federal government that result from prior actions but that will require financial resources in the future. Regardless of whether or how particular liabilities are captured on the federal balance sheet, the budget records cash flows related to those liabilities when the flows occur.

Recording Nuclear Waste Disposal Cash Flows in the Budget

Over the 1983–2006 period, the fees paid by nuclear utilities under NWPA totaled $14.8 billion, whereas expenditures totaled only $6.7 billion. As a result, the nuclear waste disposal program has reduced the cumulative net deficit (and thus the need for federal borrowing) by about $8 billion.

Because receipts to the Nuclear Waste Fund have exceeded spending, balances have grown significantly since the fund's inception, and significant amounts of interest have been credited to the fund. Intragovernmental interest is not a budgetary receipt and does not affect the size of annual deficits; however, it does add to the total amount of resources authorized to be made available for nuclear waste disposal.

Going forward, any future expenditures from the Nuclear Waste Fund in excess of annual receipts from industry fees—that is, drawing down the balances in the fund or spending the interest being credited to the fund—will increase annual deficits or reduce future surpluses. Under current budgetary procedures, all spending from the fund is considered discretionary and counts against the appropriation committees' spending allocations. Income from fees, by contrast, is recorded on the mandatory side of the budget.

Environmental and Disposal Liabilities on the Federal Balance Sheet

Along with DOE's liability for the costs of disposing of civilian nuclear waste, the federal government has substantial liabilities related to the costs of mitigating hazardous and radioactive waste that the government generated (or is required, by law or regulation, to remediate). Most of those liabilities involve the contamination of soil, water, and facilities at thousands of sites, contamination that arose from operating the nation's nuclear weapons complex. Federal financial statements prepared by the Department of the Treasury indicate that over the next 75 years, the government faces more than $300 billion in costs to clean up, dispose of, and monitor that contamination.[14]

End Notes

[1] Statement of Edward F. Sproat III, Director, Office of Civilian Radioactive Waste Management, Department of Energy, before the Subcommittee on Energy and Water Development of the House Committee on Appropriations, March 28, 2007.

[2] Data supplied to the Congressional Budget Office in September 2007 by the Department of Energy's Office of Civilian Radioactive Waste Management.

[3] Department of Energy, Office of Civilian Radioactive Waste Management, *Analysis of the Total System Life Cycle Cost of the Civilian Radioactive Waste Management Program,* DOE/RW-0533 (May 2001).

[4] Ibid.

[5] Department of Energy, Office of Civilian Radioactive Waste Management, *Nuclear Waste Fund Fee Adequacy: An Assessment,* DOE-RW-0534 (May 2001).

[6] Bechtel SAIC Company, L.L.C., *Total System Life Cycle Cost for Site Recommendation: Letter Report,* TDR-CRW-AD-000001 REV 00 (prepared for the Department of Energy, Office of Civilian Radioactive Waste Management, February 2002).

[7] Government Accountability Office, *Department of Energy: Major Construction Projects Need a Consistent Approach for Assessing Technology Readiness to Help Avoid Cost Increases and Delays,* GAO-07-336 (March 2007).

[8] Statement of Edward F. Sproat III, March 28, 2007.

[9] *Alabama Power Co.* v. *United States,* 307F.3d1300(2002).

[10] In 2006, more than 8,000 individual payments from the Judgment Fund amounted to nearly $0.7 billion; over the past 10 years, such payments have averaged around $1.2 billion annually. Most of the payments are made to settle claims related to federal employment, torts, property loss, discrimination, medical malpractice, and contract disputes.

[11] Statement of Edward F. Sproat III, March 28, 2007.

[12] Ibid.

[13] For more discussion, see Congressional Budget Office, *Comparing Budget and Accounting Measures of the Federal Government's Fiscal Condition* (December 2006).

[14] See Department of the Treasury, Financial Management Service, *2006 Financial Report of the United States Government,* available at www.fms.treas.gov/fr/06frusg/06frusg.pdf.

In: Rethinking Nuclear Power in the United States ISBN: 978-1-60692-152-4
Editor: Kenneth A. Vellis

Chapter 5

NUCLEAR POWER PLANTS: VULNERABILITY TO TERRORIST ATTACK*

Mark Holt and Anthony Andrews

SUMMARY

Protection of nuclear power plants from land-based assaults, deliberate aircraft crashes, and other terrorist acts has been a heightened national priority since the attacks of September 11, 2001. The Nuclear Regulatory Commission (NRC) has strengthened its regulations on nuclear reactor security, butcritics contend that implementation by the industry has been too slow and that further measures are needed. Several provisions to increase nuclear reactor security were included in the Energy Policy Act of 2005, signed August 8, 2005. The law requires NRC to conduct "force-on-force" security exercises at nuclear power plants at least once every three years and to revise the "design-basis threat" that nuclear plant security forces must be able to meet, among other measures.

Nuclear power plants have long been recognized as potential targets of terrorist attacks, and critics havelongquestioned the adequacy of the measures required of nuclear plant operators to defend against such attacks. Following

* This is an edited, reformatted and augmented version of a CRS Report for Congress publication dated August 2007.

the September 11, 2001, attacks on the Pentagon and the World Trade Center, the Nuclear RegulatoryCommission (NRC) began a "top-to-bottom" review of its security requirements. On February 25, 2002, the agency issued "interim compensatory security measures" to deal with the "generalized high-level threat environment" that continued to exist, and on January 7, 2003, it issued regulatory orders that tightened nuclear plant access. On April 29, 2003, NRC issued three orders to restrict security officer work hours, establish new security force training and qualification requirements, and increase the "design basis threat" (DBT) that nuclear security forces must be able to defend against.

In the Energy Policy Act of 2005 (EPACT), Congress imposed a statutory requirement on the NRC to initiate rulemaking for revising the design basis threat.[1] The DBT, a classified document, describes general characteristics of adversaries that nuclear plants and nuclear fuel cycle facilities must defend against, including radiologic sabotage and theft of strategic special nuclear material. EPACT required NRC to consider 12 factors in revising the DBT, including but not limited to an assessment of various terrorist threats, sizable explosive devices and modern weapons, attacks by persons with sophisticated knowledge of facility operations, and attacks on spent fuel shipments. NRC approved its final rule to amend the DBT (10 C.F.R. Part 73.1) on January 29, 2007.[2]

Plant Physical Security

Under the regulations in place prior to the September 11 attacks (10 C.F.R. 73 — Physical Protection of Plants and Materials), all NRC-licensed commercial nuclear power plants must have a series of physical barriers and a trained security force. The plant sites are divided into three zones: an "owner-controlled" buffer region, a "protected area," and a "vital area." Access to the protected area is restricted to a portion of plant employees and monitored visitors, with stringent access barriers. The vital area is further restricted, with additional barriers and access requirements. The security force must comply with NRC requirements on pre-hiring investigations and training.[3]

NRC published a proposed rule on October 26, 2006, to impose additional reactor security requirements.[4] It would also codify in regulation various NRC security orders issued since the 9/11 attacks and implement security provisions included in EPACT.

Design Basis Threat

The design basis threat is used by NRC licensees as the basis for implementing defensive strategies of a specific nuclear plant site through security plans, safeguards contingency plans, and guard training and qualification plans. One of NRC's April 2003 regulatory orders changed the DBT to "represent the largest reasonable threat against which a regulated private guard force should be expected to defend under existing law," according to the NRC announcement.

NRC's final rule for revising the DBT took effect April 18, 2007. As directed by the Energy Policy Act, the final rule has the principal objective of making the security requirements imposed by the April 29, 2003, DBT orders generically applicable. Although specific details of the revised DBT were not released to the public, in general the final rule

- clarifies that physical protection systems are required to protect against diversion and theft of fissile material;
- expands the assumed capabilities of adversaries to operate as oneor more teams and attack from multiple entry points;
- assumes that adversaries are willing to kill or be killed and are knowledgeable about specific target selection;
- expands the scope of vehicles that licensees must defend against to include water vehicles and land vehicles beyond four-wheel-drive type;
- revises the threat posed by an insider to be more flexible in scope; and
- adds a new mode of attack from adversaries coordinating a vehicle bomb assault with another external assault.

In 2006, the Government Accountability Office (GAO) reviewed the upgrades in nuclear plant security and found a generally logical and well-defined process. GAO found NRC staff trained in threat assessment who monitor information provided by intelligence agencies and screen the information to evaluate particular terrorist capabilities for inclusion in the DBT. However, the NRC produced a revised DBT that generally, but not always, corresponded to the threat assessment staff's original recommendations.[5]

The DBT final rule excluded aircraft attacks, which raised considerable controversy. In approving the rule, NRC rejected a petition from the Union of

Concerned Scientists to require that nuclear plants be surrounded by aircraft barriers made of steel beams and cables (the so-called "beamhenge" concept). Critics of the rule charged that deliberate aircraft crashes were a highly plausible mode of attack, given the events of 9/11. However, NRC contended that power plants were already required to mitigate the effects of aircraft crashes and that "active protection against airborne threats is addressed by other federal organizations, including the military."[6] Additional NRC action on aircraft threats is discussed below.

Force-On-Force Exercises

EPACT codified an NRC requirement that each nuclear power plant conduct security exercises every three years to test its ability to defend against the design basis threat. In these "force-on-force" exercises, monitored by NRC, an adversary forcefrom outsidetheplant attempts to penetrate the plant's vital area and damage or destroy key safety components. Participants in the tightly controlled exercises carry weapons modified to fire only blanks and laser bursts to simulate bullets, and they wear laser sensors to indicate hits. Other weapons and explosives, as well as destruction or breaching of physical security barriers, may also be simulated. While one squad of the plant's guard force is participating in a force-on-force exercise, another squad is also on duty to maintain normal plant security. Plant defenders know that a mock attack will take place sometime during a specific period of several hours, but they do not know what the attack scenario will be. Multiple attack scenarios are conducted over several days of exercises.

Full implementation of the force-on-force program began in late 2004. Standard procedures and other requirements have been developed for using the force-on-force exercises to evaluate plant security and as a basis for taking regulatory enforcement action. Many tradeoffs are necessary to make the exercises as realistic and consistent as possible without endangering participants or regular plant operations and security.

NRC required the nuclear industry to develop and train a "composite adversary force" comprising security officers from many plants to simulate terrorist attacks in the force-on-force exercises. However, in September 2004 testimony, GAO criticized the industry's selection of a security company that guards about half of U.S. nuclear plants, Wackenhut, to also provide the adversary force. In addition to raising "questions about the force's independence," GAO noted that Wackenhut had been accused of cheating on previous force-on-force exercises by the Department of Energy.[7] EPACT requires NRC to "mitigate any potential conflict of interest that could influence the results of a force-on-force exercise, as the Commission determines to be necessary and appropriate."

Emergency Response

After the 1979 accident at the Three Mile Island nuclear plant near Harrisburg, PA, Congress required that all nuclear power plants be covered by emergency plans. NRC requires that within an approximately 10-mile Emergency Planning Zone (EPZ) around each plant, the operator must maintain warning sirens and regularly conduct evacuation exercises monitored by NRC and the Federal Emergency Management Agency (FEMA). In light of the increased possibility of terrorist attacks that, if successful, could result in release of radioactive material, critics have renewed calls for expanding the EPZ to include larger population centers.

The release of radioactive iodine during a nuclear incident is a particular concern, because iodine tends to concentrate in the thyroid gland of persons exposed to it. Emergency plans in many states include distribution of iodine pills to the population within the EPZ. Taking non-radioactive iodine before exposure would prevent absorption of radioactive iodine but would afford no protection against other radioactive elements. In 2002, NRC began providing iodine pills to states requesting them for populations within the 10-mile EPZ.

Nuclear Plant Vulnerability

The major concerns in operating a nuclear power plant are controlling the nuclear chain reaction and assuring that the reactor core does not lose its coolant and "melt down" from the heat produced by the radioactive fission products within the fuel rods. U.S. plants are designed and built to prevent dispersal of radioactivity, in the event of an accident, by surrounding the reactor in a steel-reinforced concrete containment structure, which represents an intrinsic safety feature. Two major accidents have taken place in power reactors, at Three Mile Island (TMI) in 1979 and at Chernobyl in the Soviet Union in 1986. Although both accidents resulted from a combination of operator error and design flaws, TMI's containment structure effectively prevented a major release of radioactivity from a fuel meltdown caused by the loss of coolant. In the Chernobyl accident, the reactor's protective barriers were breached when an out-of-control nuclear reaction led to a fierce graphite fire that caused a significant part of the radioactive core to be blown into the atmosphere.

Vulnerability from Air Attack

Nuclear power plants were designed to withstand hurricanes, earthquakes, and other extreme events. But deliberate attacks by large airliners loaded with fuel, such as those that crashed into the World Trade Center and Pentagon, were not analyzed when design requirements for today's reactors were determined.[8] A taped interview shown September 10, 2002, on Arab TV station al-Jazeera, which contains a statement that Al Qaeda initially planned to include a nuclear plant in its 2001 attack sites, intensified concern about aircraft crashes.

In light of the possibility that an air attack might penetrate the containment building of a nuclear plant, some interest groups have suggested that such an event could be followed by a meltdown and widespread radiation exposure. Nuclear industry spokespersons have countered by pointing out that relatively small, low-lying nuclear power plants are difficult targets for attack, and have argued that penetration of the containment is unlikely, and that even if such penetration occurred it probably would not reach the reactor vessel. They suggest that a sustained fire, such as that which melted the steel support structures in the World Trade Center buildings, would be impossible unless an attacking plane penetrated the containment completely, including its fuel-bearing wings. According to former NRC Chairman Nils Diaz, NRC studies "confirm that the likelihood of both damaging the reactor core and releasing radioactivity that could affect public health and safety is low."[9]

Although NRC's final rule on the design basis threat (discussed above) excludes aircraft attacks, NRC announced on April 24, 2007, that it would issue a proposed rule requiringlicense applicants for new reactors toassess potential design improvements that would improve protection against impact by large commercial aircraft. The proposed rule would affect new reactor designs not previously certified by NRC.[10] Westinghouse submitted changes in the design of its AP1000 reactor to NRC on May 29, 2007, proposing to line the inside and outside of the reactor's concrete containment structure with steel plates to increase resistance to aircraft penetration.[11]

Spent Fuel Storage

When no longer capable of sustaining a nuclear chain reaction, "spent" nuclear fuel is removed from the reactor and stored in a pool of water in the reactor building and at some sites later transferred to dry casks on the plant grounds. Because both types of storage are located outside the reactor containment structure, particular concern has been raised about the vulnerability of spent fuel to attack by aircraft or other means. If terrorists could breach a spent fuel pool's concrete walls and drain the cooling water, the spent fuel's zirconium cladding could overheat and catch fire.

The National Academy of Sciences (NAS) released a report in April 2005 that found that "successful terrorist attacks on spent fuel pools, though difficult, are possible," and that "if an attack leads to a propagating zirconium cladding fire, it could result in the release of large amounts of radioactive material." NAS recommended that the hottest spent fuel be interspersed with cooler spent fuel to reduce the likelihood of fire, and that water-spray systems be installed to cool spent fuel if pool water were lost. The report also called for NRC to conduct more analysis of the issue and consider earlier movement of spent fuel from pools into dry storage.[12] The FY2006 Energy and Water Development Appropriations Act (P.L. 109-103, H.Rept. 109-275) provided $21 million for NRC to carry out the site-specific analyses recommended by NAS.

NRC has long contended that the potential effects of terrorist attacks are too speculative to include in environmental studies for proposed spent fuel storage and other nuclear facilities. However, the U.S. Court of Appeals for the 9th Circuit ruled in June 2006 that terrorist attacks must be included in the environmental study of a dry storage facility at California's Diablo Canyon nuclear plant. NRC reissued the Diablo Canyon study May 29, 2007, to comply with the court ruling, but it did not include terrorism in other recent environmental studies.[13]

Regulatory and Legislative Proposals

Critics of NRC's security measures have demanded both short-term regulatory changes and legislative reforms. A fundamental concern was the nature of the DBT, which critics contended should be increased to include a number of separate, coordinated attacks. Critics also pointed out that licensees are required to employ only a minimum of five security personnel on duty per plant, which they argue is not enough for the job.[14] Nuclear spokespersons responded that the actual securityforce for the nation's 65 nuclear plant sites numbers more than 5,000, an average of about 75 per site (covering multiple shifts). Nuclear plant security forces are also supposed to be aided by local law enforcement officers if an attack occurs.

Because of the growing emphasis on security, NRC established the Office of Nuclear Security and Incident Response on April 7, 2002. The office centralizes security oversight of all NRC-regulated facilities, coordinates with law enforcement and intelligence agencies, and handles emergency planning activities. Force-on-force exercises are an example of the office's responsibilities.

End Notes

[1] P.L. 109-58, Title VI, Subtitle D — Nuclear Security (Secs. 651-657). Sec. 651 adds Atomic Energy Act Sec. 170E. Design Basis Threat Rulemaking.

[2] *Federal Register*, March 19, 2007 (vol. 72, no. 52), Nuclear Regulatory Commission, Design Basis Threat, Final Rule, pp. 12705-12727.

[3] General NRC requirements for nuclear power plant security can be found in 10 C.F.R. 73.55.

[4] *Federal Register*, October 26, 2006 (vol. 71, no. 207), Nuclear Regulatory Commission, Power Reactor Security Requirements, Proposed Rule.

[5] GAO, *Nuclear Power Plants — Efforts Made to Upgrade Security, But the Nuclear Regulatory Commission's Design Basis Threat Process Should Be Improved*, GAO-06-388, Report to the Chairman, Subcommittee on National Security, Emerging Threats, and International Relations, Committee on Government Reform, House of Representatives, March 2006.

[6] NRC Office of Public Affairs, "NRC Approves Final Rule Amending Security Requirements," News Release No. 07-012, January 29, 2007.

[7] Government Accountability Office. *Nuclear Regulatory Commission: Preliminary Observations on Efforts to Improve Security at Nuclear Power Plants.* Statement of Jim Wells, Director, Natural Resources and Environment, Government Accountability Office, to the Subcommittee on National Security, Emerging Threats, and International Relations, House Committee on Government Reform. September 14, 2004. p. 14.

[8] Meserve, Richard A., NRC Chairman, "Research: Strengthening the Foundation of the Nuclear Industry," Speech to Nuclear Safety Research Conference, October 29, 2002.

[9] Letter from NRC Chairman Nils J. Diaz to Secretary of Homeland Security Tom Ridge, September 8, 2004.

[10] NRC Office of Public Affairs, "NRC Proposes Adding Plane Crash Security Assessments to New Reactor Design Certification Requirements," News Release No. 07-053, April 24, 2007.

[11] MacLachlan, Ann, "Westinghouse Changes AP1000 Design to Improve Plane Crash Resistance," *Nucleonics Week*, June 21, 2007.

[12] National Academy of Sciences, Board on Radioactive Waste Management, *Safety and Security of Commercial Spent Nuclear Fuel Storage, Public Report* (online version), released April 6, 2005.

[13] Beattie, Jeff, "NRC Takes Two Roads on Terror Review Issue," *The Energy Daily*, February 27, 2007.

[14] 10 C.F.R. 73.55 (h)(3) states: "The total number of guards, and armed, trained personnel immediately available at the facility to fulfill these response requirements shall nominally be ten (10), unless specifically required otherwise on a case bycase basis by the Commission; however, this number may not be reduced to less than five (5) guards."

In: Rethinking Nuclear Power in the United States ISBN: 978-1-60692-152-4
Editor: Kenneth A. Vellis

Chapter 6

WHAT ARE SPENT NUCLEAR FUEL AND HIGH-LEVEL RADIOACTIVE WASTE?*

United States Department of Energy

Spent nuclear fuel and high-level radioactive waste are materials from nuclear power plants and government defense programs. These materials contain highly radioactive elements, such as cesium, strontium, technetium, and neptunium. Some of these elements will remain radioactive for a few years, while others will be radioactive for millions of years.

Scientists worldwide agree that the safest way to manage these materials is to dispose of them deep underground in what is called a geologic repository.

NUCLEAR FUEL: FRESH AND SPENT

To make electricity, nuclear reactors use fuel made of solid ceramic pellets of enriched uranium that are sealed in strong metal tubes. The tubes are bundled together to form a *nuclear fuel assembly*. The uranium pellets are about the size of the tip of your little finger. Although small, the pellets release a tremendous amount of energy when used in a nuclear reactor. For example,

* This is an edited, reformatted and augmented version of a U. S. Department of Energy publication dated July 2007.

one pellet has an amount of energy equivalent to almost one ton of coal. The energy released by the nuclear fuel produces great heat, which boils water into steam. The steam turns giant turbines, which generate the electricity. About 20 percent of the nation's electricity is generated by nuclear power. After three or four years in a reactor, however, the uranium pellets are no longer efficient for producing electricity and the assembly is removed from the reactor. After removal, the entire assembly (now called *spent nuclear fuel*) is highly radioactive. All nuclear reactors produce spent fuel. Currently, there are reactors at commercial power plants, at government research facilities, and on about 40 percent of the U.S. Navy's submarines and ships. If we were to take all the spent fuel produced to date in the United States and stack it side-by-side, end-to-end, the fuel assemblies would cover an area about the size of a football field to a depth of about five yards.

When spent fuel is removed from a reactor, it is placed in a special pool of water contained in a steel-lined concrete basin.

CURRENT METHODS FOR STORING SPENT NUCLEAR FUEL

When spent fuel is first removed from a reactor, it is placed in a special pool of water contained in a steel-lined concrete basin. The water cools the spent fuel and protects workers and the public from radiation. After it has cooled considerably, some commercial power plants and government facilities move the fuel to dry-storage containers made of steel and/or concrete to shield radiation. The containers are either placed upright on concrete pads, or stored horizontally in metal canisters in concrete bunkers.

HIGH-LEVEL RADIOACTIVE WASTE

With the end of the Cold War, the United States has been working to close and clean up obsolete weapons plants and dispose of nuclear weapons materials. This has created a need to dispose of highly radioactive material associated with weapons production. This material is called *high-level radioactive waste.* Until the late 1970s, the United States acquired materials for nuclear weapons by reprocessing spent nuclear fuel from government-owned nuclear reactors. Reprocessing is a method of chemically treating spent fuel to separate out uranium and plutonium. The byproduct of reprocessing is a highly radioactive sludge-like residue. The Department of Energy will not ship such residue to a repository until it is solidified and sealed in stainless steel canisters.[*] All high-level nuclear waste is currently stored at government facilities.

A LONG-TERM ENVIRONMENTAL PROBLEM

Nuclear waste must be properly managed to minimize risk to the environment and to the health and safety of future generations. Since the mid-1940s, spent nuclear fuel and high-level radioactive waste have accumulated throughout the country. Currently, they are stored in temporary facilities at some 121 sites in 39 states. These storage sites are located in a mixture of urban, suburban, and rural environments — most are located near large bodies of water. In the United States today, over 161 million people reside within 75 miles of temporarily stored nuclear waste. Current storage methods shield any harmful radiation and are presently safe. However, modern aboveground storage structures are designed for temporary storage only, and will not withstand rain, wind, and other environmental factors for the tens of thousands of years during which the waste will be hazardous.

PERMANENT DISPOSAL OPTIONS

For decades, experts throughout the world have studied many options for permanently disposing of nuclear waste — including:

- Leaving the material at current storage sites
- Burying it in the ocean floor
- Putting it in polar ice sheets
- Sending it into outer space
- Placing it deep underground in a geologic repository

After analyzing these options, most scientists agree that disposal in an underground repository is the best long-term solution for safely managing highly radioactive wastes. This opinion is reflected in a 1990 report from the National Research Council of the National Academy of Sciences, which states that there is "*a worldwide scientific consensus that deep geological disposal, the approach being followed by the United States, is the best option for disposing of highly radioactive waste.*" In 2001, the National Research Council again reaffirmed that conclusion: "*Geologic disposal remains the only scientifically and technically credible long-term solution available to meet the need for safety without reliance on active management.*" [**] As long as nuclear waste remains in a solid form and is properly shielded, it will not harm people or contaminate the environment — and over time it produces less and less radiation. The idea behind deep geologic disposal, therefore, is to keep the waste as dry and isolated as possible, for as long as possible, so that its radiation can diminish to safe levels. Isolated in a deep underground repository, the waste would not be subject to the many environmental factors that on the earth's surface would cause it to break down into radioactive particles that could be dispersed by air or water into the accessible environment.

NUCLEAR WASTE: THE FACTS

Spent nuclear fuel and high-level radioactive waste present a challenge society must address. The Department of Energy bases its plans for a proposed repository at the Yucca Mountain site on several key facts about the radioactive materials that would go into it. Among these facts are the following:

Key Fact: The wastes DOE plans to put in the proposed Yucca Mountain Repository are *solid* (not liquid).

The nuclear waste destined for disposal at a repository will be in the form of solid metals, ceramics, and glass with small amounts of radioactive gases.

Key Fact: Spent nuclear fuel and high-level radioactive waste cannot cause an explosion.

Spent nuclear fuel and high-level radioactive waste are ***not*** explosive. Even if these materials were involved in an explosion (like a transportation accident involving an oil tanker), it is physically impossible for them to cause a nuclear chain reaction like a nuclear bomb.

Key Fact: Spent nuclear fuel and high-level radioactive waste are *not* flammable.

Since these materials are composed of metals, ceramics, and glass, they cannot fuel a fire.

End Notes

[*] In the U.S., all nuclear waste must be in solid form before it can be shipped to a repository.

[**] Board on Radioactive Waste Management, *Rethinking High-Level Radioactive Waste Disposal*, National Academy Press: Washington, D.C., 1990, p. vii. And: *Disposition of High-Level Waste and Spent Nuclear Fuel: Continuing Societal and Technical Challenges*, National Academy Press: Washington, D.C., 2001, p.

In: Rethinking Nuclear Power in the United States ISBN: 978-1-60692-152-4
Editor: Kenneth A. Vellis

Chapter 7

OVERVIEW: YUCCA MOUNTAIN PROJECT[*]

United States Department of Energy

Some of the electricity used in homes across our nation is made with nuclear fuel in reactors at power plants. When the fuel that is used to make electricity at the power plants can no longer produce electricity efficiently, it is removed from the reactors. This used fuel is called spent fuel. Some nuclear materials used for national defense create waste. This waste is called high-level radioactive waste. These highly radioactive materials must be protected so they won't harm people or the environment.

The basic idea of geologic disposal is to place carefully packaged radioactive materials in tunnels deep underground.

---

* This is an edited, reformatted and augmented version of a U. S. Department of Energy publication dated November 2003.

WHAT SHOULD WE DO WITH RADIOACTIVE WASTE TO PROTECT PEOPLE AND THE ENVIRONMENT?

Scientists have studied different ways to dispose of this radioactive material. Currently, it is stored at government sites and at the power plants where it is made. However, a permanent disposal place is needed. Most scientists around the world agree that the best place to put this radioactive material is in a facility deep underground. This type of facility is called a geologic repository. The U.S. Department of Energy (DOE) has studied Yucca Mountain for more than two decades. In 2002, the secretary of energy and then the president recommended the site as suitable for a geologic repository. Congress subsequently supported this recommendation, officially designating Yucca Mountain for a proposed repository. Now the DOE can submit a license application to the Nuclear Regulatory Commission, which has three to four years to decide whether to grant a license to construct a repository.

This artist's drawing shows what a repository might look like inside Yucca Mountain, if one were to be built there. Construction would take place through tunnels underground.

WHY YUCCA MOUNTAIN?

Yucca Mountain is about 100 miles northwest of Las Vegas, Nevada, on land owned by the federal government. No one lives on Yucca Mountain. The area has a very dry climate — receiving a combined average of about 7.5 inches of precipitation per year. Approximately 95% of this total either runs off, evaporates, or is taken up by the desert vegetation. Yucca Mountain has a very deep water table. If a repository is built at Yucca Mountain, it would be located about 1,000 feet below the surface and 1,000 feet above the water table. So any water that does not run off or evaporate at the surface would have to move down nearly 1,000 feet before reaching the repository and then another 1,000 feet before it reached the water table. The dry climate is an important feature because water is the primary way by which radioactive material could move from a repository.

HOW WOULD A REPOSITORY WORK?

The basic idea of geologic disposal is to place carefully packaged radioactive materials in tunnels deep underground. This method relies on a series of barriers that prevent or slow the movement of radioactive materials from a repository. These barriers include natural ones, such as thick unsaturated rock, and man-made, or engineered, ones. These barriers also would greatly reduce the total amount of any radioactivity that could eventually reach the water table where people might pump it from the ground and use it. The current design for the potential repository calls for spent nuclear fuel and high-level radioactive waste to travel to Yucca Mountain by truck or rail in specially designed, shielded shipping containers. Once these materials arrive at the repository, they would be removed from the shipping containers and placed in double-layered, corrosion-resistant packages for burying underground. Special rail cars would carry them underground, and remotely controlled equipment would place them on supports in an underground tunnel. Once the materials are placed in the repository, scientists would continue to check that everything is working the way it should during the performance confirmation period.

Who will Determine if a Repository is Built at Yucca Mountain?

Before the DOE could construct a geologic repository and begin waste emplacement, the Department must submit a license application, go through a multi-year review and public hearing process, and then receive a construction authorization from the Nuclear Regulatory Commission (NRC). The hearing process would focus on public health and safety. Along with the review process, the hearing process is expected to take a minimum of three years after the DOE submits a license application. If the DOE receives a construction authorization, it would have to complete initial construction, and apply for and receive a license from the NRC before any waste could be received or emplaced. This includes demonstrating to the NRC that there is a reasonable expectation a repository designed for the Yucca Mountain site could protect health and safety for 10,000 years after the repository is closed.

CHAPTER SOURCES

The following chapters have been previously published:

Chapter 1 – This is an edited, reformatted and augmented version of a Congressional Budget Offie report, Publication No. 2986, dated May 2008.

Chapter 2 - This is an edited, reformatted and augmented version of a Congressional Research Service publication, Order Code RL33442, updated March 9, 2007.

Chapter 3 – These remarks were delivered as testimony given on November 14, 2007. Peter R. Orszag, Director, Congressional Budget Office, before the Committee on Energy and Natural Resources, United States Senate.

Chapter 4 - These remarks were delivered as testimony given on October 4, 2007. Kim Cawley, Chief, Natural and Physical Resources, Cost Estimates Unit, before the Committee on the Budget, United States House of Representatives.

Chapter 5 – This is an edited, reformatted and augmented version of a Congressional Research Service publication, Order Code RS21131, upated August 8, 2007.

Chapter 6 - This is an edited, reformatted and augmented version of a United States Department of Energy, Office of Civilian Radioactive Waste Management report, DOE/YMP-0338, dated July 2007.

Chapter 7 - This is an edited, reformatted and augmented version of a United States Department of Energy, Office of Civilian Radioactive Waste Management report, DOE/YMP-0026, dated November 2003.

INDEX

A

B

C

D

E

F

G

H

M

N

O

P

R

S

T

U

V

W

Y

Z

Renewable Energy: Research, Developmen[t] and Policies

Renewable Fuel Standard Issues
Daniel T. Crowe (Editor)
2009. ISBN: 978-1-60692-289-7

Ethanol and Biofuels: Production, Standards and Potential
Wesley P. Leland (Editor)
2009. ISBN: 978-1-60692-224-8
2009 ISBN: 978-1-60876-592-8 (E-book)

Wind Power: Technology, Economics and Policies
Cedrick N. Osphey (Editor)
2009. ISBN :978-1-60692-323-8

Solar Collectors: Energy Conservation, Design and Applications
Arthur V. Killian (Editor)
2009. ISBN: 978-1-60741-069-0
2009. ISBN: 978-1-60876-920-9 (E-book)

Wind Energy in Electricity Markets with High Wind Penetration
Julio Usaola and Edgardo D. Castronuovo
2009. ISBN: 978-1-60741-153-6

Renewable Energy Grid Integration: The Business of Photovoltaics
Marco H. Balderas (Editor)
2009. ISBN: 978-1-60741-324-0

Biomass Gasification: Chemistry, Processes and Applications
Jean-Pierre Badeau and Albrecht Levi (Editors)
2009. ISBN: 978-1-60741-461-2

Renewable Energies: Feasibilit[y] Time and Cost Options
John O'M. Bockris
2009. ISBN: 978-1-60876-006-0

Clean Energy: An Exporter's Guide to India and China
Isaac P. Luttrell (Editors)
2009. ISBN: 978-1-60741-329-5

Physics of Nanostructured Solar Cells
Viorel Badescu and *Marius Paules[cu]* *(Editors)*
2010. ISBN: 978-1-60876-110-4

Clean Coal
Klaes G. Douwe (Editor)
2010. ISBN: 978-1-60741-358-5

Rethinking Nuclear Power in the United States
Kenneth A. Vellis
2010. ISBN: 978-1-60692-152-4

Biological Barriers to Cellulosi[c] Ethanol
Ernest V. Burkheisser
2010. ISBN: 978-1-60692-203-3

Solar America: How, What and When?
Nash M. Perales (Editor)
2010. ISBN: 978-1-60741-333-2

Renewable Energy Grid Integration: Building and Assessme[nt]
Thomas Georgiadis (Editor)
2010. ISBN: 978-1-60741-326-4

RENEWABLE ENERGY: RESEARCH, DEVELOPMENT AND POLICIES

RETHINKING NUCLEAR POWER IN THE UNITED STATES